THE EDGE OF MEDICINE

The Technology That Will Change Our Lives

William Hanson, M.D.

palgrave
macmillan

First published in 2008 by PALGRAVE MACMILLAN® in the United States—a
division of St. Martin's Press LLC, 175 Fifth Avenue, New York, NY 10010.

Where this book is distributed in the UK, Europe and the rest of the world, this is by
Palgrave Macmillan, a division of Macmillan Publishers Limited, registered in
England, company number 785998, of Houndmills, Basingstoke, Hampshire RG21
6XS.

Palgrave Macmillan is the global academic imprint of the above companies and has
companies and representatives throughout the world.

Palgrave® and Macmillan® are registered trademarks in the United States, the United
Kingdom, Europe and other countries.

ISBN-13: 978–0–230–60575–6
ISBN-10: 0–230–60575–3

Hanson, William, M.D.
    The edge of medicine : the technology that will change our lives / William
Hanson.
        p. ; cm.
    Includes bibliographical references and index.
    ISBN-13: 978–0–230–60575–6
    ISBN-10: 0–230–60575–3
    1. Medical innovations. 2. Medical technology—Forecasting. I. Title.
[DNLM: 1. Biomedical Technology—trends. W 82 H251e 2008]
RA418.5.M4H36   2008
610.28—dc22

                                                                                    2008023688

A catalogue record of the book is available from the British Library.

Design by Letra Libre, Inc.

First edition: October 2008
10 9 8 7 6 5 4 3 2 1
Printed in the United States of America.

CONTENTS

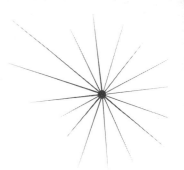

ACKNOWLEDGMENTS

T he idea for this book actually evolved from an unsuccessful concept for another book, one that, to be fair, would have been much less fun to write. Fortunately, while she passed on that first idea, Airie Stuart, senior vice president and publisher at Palgrave Macmillan, took the time, as she describes the way she sees her job, "to reshape and develop the proposal with an eye toward the larger trajectory of an author's career." After gently telling me on a conference call with my agent, Eric Lupfer from the William Morris Agency, that a book about health care's stupidities, however good, would be a downer, she suggested that I send her an alternative proposal for a book about health care's promises. I am grateful to Airie for her philosophy and to Eric for his thoughtful stewardship throughout the process. My editor, Luba Ostashevsky, has helped me to do what my sons' teachers would describe as "my best work," using, variously, blunt talk, patience, curiosity, humor and a light touch on the reins.

My siblings Chris, John, Beth and Ellen have all been cheerleaders along the way, and Beth, in particular, has acted as a rational sounding-board at every step. While several people helped focus how I have thought about the material, a few deserve particular mention. Richard Perlman and Jim Price were generous enough to lend me some of their boundless optimism and enthusiasm early on as I began, and Meg Davis provided some timely observations during the home stretch. I have also been fortunate enough to work with a number of bright, passionate, far-seeing people at several great academic institutions over the years.

I owe my greatest debt to the four people besides myself who have lived most closely with, been most impacted by and been the most enthusiastic about the project: my wife, Beth, and my three sons, Addison, Watson and Callaghan. Appropriately, one of the themes that cropped up repeatedly as I researched material is the way that strong families anchor and nourish their individual parts, which brings me to one last debt—the one I owe to my parents. I've made an inadequate attempt to discharge this debt, the most fundamental one, in the closing chapter.

INTRODUCTION

It is impossible to spend any time observing the world without concluding that curiosity is common to both humans and animals. Where we once believed that the use of tools was an exclusively human activity, we now know that various animal species manufacture and use tools in a variety of ways. The practice of medicine, as far as we know, is one of the oldest exclusively human industries. Although we don't know precisely when, at some point primate social activities such as mutual grooming and nit-picking transitioned into the beginnings of early human medicine. The first medical tools people used were their senses of sight, touch, smell, taste and hearing. And early humans weren't just observant; they acted on their observations.

Trepanation, a medical procedure in which a hole is bored in the skull, is typically used today to drain collections of blood from around the brain. Archeologists have found trepanned holes in skulls dating back to the Stone Age. We don't know what the indications were for the procedure back then, but some of the skulls were fractured, suggesting that the procedure may have been performed to treat brain injuries. Some of the trepanned Stone Age skulls show evidence of bone that healed, indicating that the patients survived the surgery in many instances.

The father of medicine, Hippocrates, diagnosed diabetes mellitus based on a patient's complaints of thirst, hunger and frequent urination. In fact, the word diabetes derives from the Greek word for siphon, because according to ancient Greek physicians, diabetics passed water like a siphon. Using his own senses as a diagnostic tool, Hippocrates tasted the urine to see if it was sweet.

One way to think of the five human senses is as dimensions of sensation. The loss of one sense would, then, be analogous to going from a five- to a four-dimensional world. The ability to see gives us the capacity to have a bird's-eye view of the lower-dimensional world, and to readily look for and find things in an area that we'd otherwise be forced to explore in a slower, less efficient fashion. Without sight, our human explorations could be confined to feeling around with our hands and listening for sound cues—which is actually a pretty good analogy for what we sometimes do in medical research: We grope around in a systematic fashion for new knowledge and new treatments using the research tools we have available in a given era.

Every once in a while, however, a new research technology comes along that effectively gives us a new sense with which to explore our world, and a burst of medical advances ensues. For example, Anton van Leeuwenhoek developed powerful microscope lenses in the mid-1600s and used them to systematically explore the world around him, thereby discovering what amounted to a new world. He put everything he could think of under the microscope and discovered bacteria, protozoa, even spermatozoa on what must have been a very surprising day.

The microscope lens augmented our ability to see and thereby extended our diagnostic capabilities. Today's CT and MRI scans are the remote descendants of the lens. In fact, the lineage of many of today's medical instruments can be traced to antiquity. Hippocrates used his sense of taste as a tool, and he also smelled diseases. He described the smells of liver and kidney diseases on the breath of his patients; in my lab today, we are working with an "electronic nose" to identify the smells of pneumonia, sinusitis and cancer on the breath of our patients.

Leeuwenhoek's microscope revealed a whole new, previously unsuspected world to scientists of his time, one that was so astonishing that many of them didn't believe his observations at first. Lenses of one sort or another have now been developed to see out into the universe and down to the individual atom. We have actually developed tools to shatter the atom in our relentless search for the most fundamental particles. In fact, the first chapter of this book will describe the lengths we have gone to in order to find the elusive Higgs boson—a critical missing piece to the model physicists currently believe best explains the nature of time and space. The pursuit of fundamental particles has already paid huge medical dividends. In fact scientists and doctors have found ways to conscript advances in almost every area of human investigation for the service of medicine; and we now have more tools than ever before to do something about what we find.

We have tamed the atom and used it for radiation-based treatments. We've designed molecular machines for medical diagnosis and treatment. We've enlisted robots and smart computers in the war on disease. We've identified the cellular equivalents of Adam and Eve. We're now able to look at and manipulate human DNA—the machine inside the man; and we're decoding the human genome.

Today, we are entering an era in which genetic information, stem cells and nanomolecular engineering will transform the world of medicine, providing us with many new dimensions of data and treatment. In this book we'll explore the evolution of medicine from past to present, showing how we got to where we are, and where we're headed in the future.

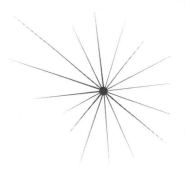

CHAPTER 1

DEUS EX MACHINA

My father was a physician whose career spanned the four decades from the early 1950s to the early 1990s, and many things changed during his lifetime. For example, in the 1950s, the RJ Reynolds Tobacco Company actually ran a campaign with the slogan "More doctors smoke Camels than any other cigarette," and many of the pictures in my father's medical school yearbook show him and classmates with cigarettes in hand. However, the really big changes occurred during the latter part of his career (when my own career was just beginning), during which medicine began to evolve from a mom-and-pop, cottage industry into the highly competitive, rapidly advancing, multinational *business* it is today.

In the late 1970s and early 1980s, I worked at the same medical center that I do now, in an office that was then called Data Processing—the hospital division that managed patient's bills and accounts payable. There was exactly one computer in the entire hospital—in the basement—and it dined exclusively on IBM punch cards (those heavy, rectangular paper cards formerly used for census-taking, school examination registration, time cards and, perhaps most notoriously, for voting in Florida during the U.S. presidential election in 2000—the piece punched out of the card is, of course, a chad). Punch cards were used to keep track of the medical bills. There were data entry clerks who punch-typed the charges onto the cards, which were then sorted into piles secured with a rubber band. Periodically, the computer would be fed a pile of punch cards upon which it would chew noisily for some time and eventually spit out a ream of data on a teletype machine.

I was the hospital's only data analyst, and in order to get lunch every day, I walked down a creaky corridor lined with cases of memorabilia, past the administrative offices of the hospital and down the steps into the cafeteria. The hospital library was one flight up those same stairs, and the medical staff would go there to pass some leisure time browsing through medical journals. Less than ten people worked in hospital administration—the chief operating officer, the chief financial officer, the chief nursing officer and a few support people. It was the job of the folks in this office to keep the books, buy things that were needed and make sure the bills got paid—nothing more. The concept of engaging in overt competition to attract patients was inconceivable: The patients just came to us, and the hospital's officers did pretty much what they were told by the doctors who sat on the hospital's board.

As I drove to work in the early 1980s, I passed plenty of billboards, none of which advertised hospitals or doctors. In fact, the thought of self-promotion was abhorrent to the medical profession at that point. There were no signs on the city buses asking "Have you been misdiagnosed?" There were very few regulatory agencies, and magazines didn't publish lists of the best doctors and hospitals. Medical benefits were a little employee perk in the same category as parking and free company business cards.

In short, medicine back then was a pretty sleepy, gentlemanly affair, and some hospitals were like the fat, slow-moving dodos from the island of Mauritius. Today, however, a mere 25 years later, we are in a medical evolutionary arms race. Computer chips are ubiquitous—there are probably 20 in my office alone, what with desktops, laptops, telephones, cell phones, a camera and other gadgets. Data analysts have found their own ecological niche; and 30 *percent* of health care workers are administrators. In fact, upwards of 30 percent of health care dollars go toward medical administration. Hospitals now have powerful CEOs, many of whom have advanced degrees in health care administration, an educational track that teaches how to control costs, grow product lines and capture market share.

Successful hospitals rank highly on performance and patient satisfaction scores and garner certificates of excellence. Some of the certifying organizations could be described as parasitic: They grow their business *de novo* by defining excellence in some new way, trademarking their brand and then, for a fee, determining whether individual hospitals meet their standards. Malpractice lawyers are everywhere, having found a vast new food source, and rummage around largely untrammeled because their fellow lawyers tend to make the laws. In the medical profession, we have our own lawyers to make sure we stay current with rapidly changing laws and regulations. Looking back, it's as if an alien ship landed on the medical world of thirty years ago, bringing with it innovation as well as unintended disease, and things began to change almost immediately thereafter for better and for worse.

Every city and country still has its medical dodo hospitals, but medical evolution has proceeded quite rapidly where environmental changes have compelled change. The development of specialty hospitals (such as centers for the treatment of obesity in the United States) and private hospitals in countries with public health systems are but two examples of successful evolutionary adaptations by doctors and hospitals in resource-constrained environments. Of course today's profitable obesity center may become tomorrow's dodo hospital due to changes in regulations and reimbursement or, ideally, to the disappearance of obesity as a problem.

The current landscape of medicine and its place in the larger society is, at best, very confusing when viewed from the ground. We are in what Clausewitz described as the fog of war. On the one hand we hear a steady stream of dire predictions about the all-engorging growth of health care in every modern country in the world, while on the other, there is a competing flow of information about tantalizing new therapies, such as proton beams, that can cure or extend life. Some of the latter will represent major breakthroughs while others will turn out to be no better than or even worse than their less costly predecessors.

For centuries, medical knowledge was as closely held as the secrets to magic acts, but today we suffer from a surfeit of electronic information that often leaves us more confused than before. The biggest problem right now, both for those of us who treat medical problems and for our patients, not to mention policy makers and payers, is that things are changing so rapidly.

Proton beam therapy is a perfect example of the technologies and treatments that are at the leading edge of medicine and that you'll find in the hospital of the future. Developed by physicists, in their search for fundamental particles, this therapy is technologically sophisticated; even as a critical element of a treatment assembly line, it can be programmed to give treatments precisely calibrated to the specific needs of each individual patient.

Physicists and physicians are both engaged in a seemingly relentless quest to build bigger machines to interact with ever-smaller targets with greater precision; the two fields are inextricably interwoven. Nobel Prize–winning physicist Marie Curie discovered radioactive elements and forms of radiation that are fundamental to the practice of medicine; this same radiation killed both Marie and her equally talented daughter. In 1959, Nobel Prize winner Richard Feynman described nanotechnology and nanoscale medical devices that are now in the process of revolutionizing medicine. The World Wide Web was originally created as a tool for interaction among particle physicists at the European Council for Nuclear Research, commonly known as CERN; the Web is now, among many other things, a wide-reaching vehicle for the dissemination and practice of medicine, as we'll see in a later chapter. Another extraordinary physicist, Stephen Hawking, has amyotrophic lateral sclerosis; it is only by virtue of advances in the treatment of that disease that he is able to continue using his energies to further our understanding of arcane concepts pertaining to black holes and the nature of space and time.

The Higgs boson is the elusive, highly sought after, as yet theoretical fundamental particle that physicists desperately need in order to tidy up one of their theories of how all things work. The boson is the last unobserved member of the particle family belonging to the standard model of physics. Physicists describe it as a rumor crossing a crowded room because it, too, causes transient clustering and massing in the wake of its passage. In essence, the boson is believed to be the fundamental particle that gives all other things—planets, people and protons—mass.

If a large hadron collider, or LHC, sounds to you like something out of a science fiction movie, you're not too far off. It is actually the gigantic nuclear particle accelerator and collider located outside of Geneva, Switzerland, at CERN, with which modern physicists plan to find the Higgs boson, using what amounts to an enormous ray gun. The LHC was built with the collabo-

ration of more than two thousand scientists and hundreds of separate universities and took ten years to construct. The collider and its associated particle de-
tectors, electron magnets and laboratories are housed in a 26.5-kilometer-long
tunnel that is a little less than 4 meters in diameter and crosses the serpentine
border between France and Switzerland several times. It lies at average depths
of between 50 and 175 meters underground to minimize its impact on the environment and to prevent harmful radiation exposure to people walking on the
land above. The discoveries made within that hidden tunnel will likely change
our world above it irreversibly.

This giant machine is designed to fire one beam of protons traveling
clockwise at another traveling counterclockwise—both beams at almost the
speed of light—in hopes that information about debris from the resulting col-
lision will explain some of the still-unanswered fundamental questions in
physics. This a time honored way of finding things out that was invented by
and has been practiced for centuries by boys. Fortunately, the proton is a generally well-behaved member of the hadron family (a class of particles com-
posed of quarks), which also includes neutrons and mesons, and the large
hadron collider is based on the design of older linear particle accelerators, or
atom smashers, upon which, in turn, some of today's medical radiation devices
are modeled.

Over 1,600 superconducting magnets cooled with liquid helium are used
to accelerate and steer the proton beams in the LHC. The energy developed in
each of the circling proton beams is equivalent to that of a high-speed train
traveling at 150 kilometers an hour. When the two beams of protons collide,
an unimaginable amount of energy is released—in fact, the temperature at the
contact point is a hundred thousand times hotter than the center of the sun.
On the other hand, the cooling system for the magnets in the tunnel—the
cryogenic distribution system—keeps parts of the collider at temperatures
lower than that of outer space; we'll see in a later chapter how our understanding of cryonics may one day allow us to travel to outer space.

The protons in the beams make over 11,000 circuits of the tunnel every
second, and the attendant forces are so powerful that there is at least a theoretical possibility that the proton collisions will produce microscopic black holes.
This sounds really ominous but physicists reassure us that, if formed, these

small black holes will deflate, like balloons, by blowing off energy through a process called Hawking radiation, named after the aforementioned physicist Stephen Hawking.

Hadron, or proton beam therapy describes the use of a beam of particles, exactly like the ones flying around the 26.5-kilometer hadron racecourse in Switzerland, to treat cancer in humans. Proton beam therapy is only available in a few places in the world. It differs from traditional radiation therapy in its use of beams of particles rather than x-ray waves for treatment. Unlike x-rays that deliver radiation to all the tissues along the path of the beam, causing a little bit of damage all along the way, proton beams pass harmlessly through the skin and overlying tissues to deliver their radiation into the target without injuring the surrounding tissues.

The first suggestion that protons might be used medically to treat tumors came from Robert R. Wilson, one of the physicists who worked on the Manhattan project with Albert Einstein, Richard Feynman and Robert Oppenheimer. A multidimensional scientist like Feynman (who sketched, painted, kept an office in a topless bar and played practical jokes), Wilson was also a sculptor, human rights advocate and a little bit of a rebel. Wilson described the potential for proton therapy in 1946. Having observed that proton beams fired from an accelerator give off a burst of radiation just before they come to a stop, he realized that a beam could be tuned to deliver energy to a very precise area, even one deep within the body.

Wilson proposed that bursts of protons could be trained, like horses, to gallop up to a tumor in formation and then stop on a dime (or on a pinpoint in the case of protons) to deliver their payload of radiation. It is instructive to contrast this elegant medical treatment with what happens in the hadron collider in Switzerland, where two columns of protons are smashed violently into one another at nearly the speed of light, shattering what we once thought were *the* fundamental particles—neutrons, protons and electrons— into still-smaller particles with names such as charm quark, strange quark, muon, gluon and baryon. In fact, all of the early proton treatments for medical disease were performed in particle accelerators originally built for physics research, such as the one at the Harvard Cyclotron Laboratory, which Wilson helped design, as well as devices in the Soviet Union, Switzerland, Japan and Sweden. The first accelerator built specifically for medical therapy wasn't constructed until 1988.

Highly targeted proton therapy is particularly useful when a tumor lies in close proximity to critical nerves or organs, such as the brain, prostate, esophagus, lungs and eyes. Because of the proximity of the tumor to its neighbors, the dose of radiation that can be safely administered is limited by the potential for unacceptable injury to critical normal tissue nearby. Proton beams act like smart weapons. The oncologist uses the coordinates of the tumor, based on imaging data from a CT or MRI scan, to design a course of radiation in which the beam is shaped to conform to the silhouette of the lesion. The protons are then energized to the exact level needed to deliver their radiation precisely into the tumor.

I recently had the opportunity to walk around a proton beam accelerator under construction at the Roberts Proton Therapy Center, in the Perelman Center for Advanced Medicine at the Hospital of the University of Pennsylvania. When complete, the center will be part of the largest proton therapy institute in the world, and after having seen the innards of the accelerator before the walls were installed, I can see why this $150 million piece of machinery is called the world's most expensive and complex medical device. While Geneva's large hadron collider is the world's largest *scientific* instrument, designed to answer big fundamental questions about the universe, its medical counterpart is a worthy little sibling.

The proton therapy cyclotron in the Roberts Center is a 220-ton device at the heart of a football-field-sized building, part of a beautiful medical center designed by the architect Rafael Viñoly. Viñoly specializes in the design of seemingly gravity-defying buildings such as the Frederick P. Rose Hall, home of Jazz at Lincoln Center in New York, the Kimmel Center for the Performing Arts in Philadelphia, Princeton's Carl Icahn Laboratory of the Lewis-Sigler Institute for Integrative Genomics and a lattice-work concept for the World Trade Center site, from which two beams of white light vanish into the night sky.

The proton beam generated in the Roberts Center cyclotron can be split into five separate independent sub-beams that are directed to patients by a series of bending, focusing and routing electromagnets. Each of the five beams is delivered into separate suites, multiplying the patient treatment capacity of the center accordingly. The treatment rooms have walls 18 feet thick, like the equipment silos of the hadron tunnel, to contain stray neutron radiation. The building itself, like the CERN collider, is mostly underground, hidden well below the airy superstructure.

At one point during my tour around the scaffolding and curing concrete, I stood at the edge of a doorway, blocked by a flimsy, temporary guardrail, that opened into space halfway up the wall of a cavernous room that will eventually be one of the treatment suites. As I watched, construction workers were welding together a 35-foot high, ninety-ton metal cage known as a gantry, capable of rotating 360 degrees around the patient. The proton beam is guided from the cyclotron and then around this gantry, through a nozzle, past the skin and into the cancer with a 20,000-pound magnet; it is shaped just prior to delivery to conform to the exact silhouette of the patient's tumor. As I stood there with my toes hanging over the sill, looking up one story to the top and down another to the bottom, I was struck by the disproportion between this gigantic machine and the relatively tiny pallet at its center on which the patient will lie during treatment.

When the Roberts Center is finished, that doorway into space will appear to be an ordinary entrance to a functional room and the gantry won't be visible. Patients will be unaware of the surrounding machinery during a course of treatment because it will be hidden behind the walls. And patients will barely have time to think about what's happening around them because individual treatments will be short. Still, even a cursory understanding of the events going on during their time there will leave patients with a good measure of shock and awe at the current breadth of our knowledge of physics, human biology and the lengths to which we will go to fix a body when it's broken.

In order to fully appreciate what an enormous human achievement proton beam therapy represents, it's instructive to follow what happens in a stepwise fashion between the time hydrogen gas enters the system at one end and the point at which, on our command, a rank of prancing protons comes to a dead stop in the middle of a cancer tumor hundreds of feet away. That tumor may only be a few *millimeters* large and located in a patient's eye, such as a two-millimeter melanoma on an otherwise blue iris. Step one is the injection of hydrogen atoms into an electrical field, in which the single electron is stripped off leaving a free, positively charged proton. In the next step, a bunch of these protons are spun around in a cyclotron ten million times a second and are given a little electrical nudge with each rotation, like children on a playground merry-go-round, until they reach an energy level of millions of electron volts. Bursts of these energized, charged particles then exit the cyclotron and are channeled along a virtual proton road to the exact spot that we want them to park, so to speak, to discharge their accumulated energy in a blast and consequently rip electrons right out of the orbits of the molecules making up the cancer cells. And the protons don't just tear off the cellular equivalent of a bit of paint or a side-view mirror; they come in hard, right into the front end of a cancer cell and destroy its DNA. In fact, one of the advantages of proton therapy over standard x-rays is that the proton beam's radiation so thoroughly ravages the tumor's DNA that its ordinary repair mechanisms are completely ineffective.

It is important to understand that the proton therapy beam is a safe form of treatment because its energy is very carefully controlled. If you trained the much more powerful proton beam from the large hadron collider on a brain tumor in a patient, it would turn that whole patient's head into something called quark-gluon plasma, which would, for a whole bunch of reasons, be very awkward to have to explain to a plaintiff-friendly jury in the event of a malpractice suit. The energy of the LHC beam is measured in tera electron volts, or a million times the energy used in medical proton beam treatments.

Prometheus gave us the gift of fire a long time ago; but, for better and for worse, we've now found a way, on our own, to play with nuclear matches. We humans can now confidently harness the enormous amounts of energy stored in atoms to answer deep questions about the universe, to create unimaginably powerful bombs or to shred the molecular machinery of a cancer cell.

Patients treated with proton therapy will include children, who are at greatest risk from the long-term effects of traditional radiation, and adults with a variety of localized cancers that haven't yet spread. To date, the largest population of patients undergoing proton treatment are those with prostate cancer, a disease for which there are already a bewildering variety of alternative treatments. However, proton beam irradiation has major advantages over other forms of prostate treatment: It doesn't require surgery and it can potentially avoid the worst side effects of traditional x-ray-based radiation and surgery, such as impotence and incontinence, by sparing the nerves, bladder and rectum around the prostate.

Proton beam therapy can be characterized as a medical tour de force, in which the treatment is delivered like a *deus ex machina*. The patient walks into a room, lies down on a bed, and, for the few minutes he's there, all of this magical stuff happens around him and to him, painlessly, silently, perhaps while he listens to his iPod. Then he gets up and leaves after paying his co-pay. Although technically an outpatient treatment, it exemplifies the kind of scientifically sophisticated methods that advanced medical centers in the United States, Europe and Asia are deploying in next-generation facilities that are definitely not your father's hospital, and not my father's either. The hospitals of the future will be technologically rich, patient-centered facilities employing cutting-edge electronics, robotic attendants, smart computer software and personalized treatments precisely tailored to each patient, such as gene and stem cell therapy.

The moment a patient enters the hospital door, her stay will be tracked electronically so we'll always know where she is, how long she's been waiting and where she's due to go next. Robots or electronic maps will lead visitors through the hospital and they won't have to stumble around the often-bewildering mazes of buildings and offices as they do today. Families waiting for the patient who disappeared into an operating room will soon wander freely around the hospital carrying something like those restaurant pagers that light up and vibrate when your table is ready—only this one will be used to summon a patient's husband back to talk to the surgeon at the end of the operation or a patient who has been waiting to see the doctor. Perhaps we'll even find a device that will keep the doctor on schedule.

Robotic patient and equipment transport will become routine as the technology matures and as manpower costs continue to escalate. Overhead hospital communications, like "Dr. Jones, please call Radiology," are already becoming a thing of the past and will vanish as automated communications using smart devices supersede pagers and loudspeakers. Communication badges have already been developed that allow a doctor to call a colleague by saying something like "contact on-call hematologist." These systems use voice recognition and smart software to identify that on-call hematologist and contact her directly, bypassing the inefficient operator-based methods used in most hospitals today. Critical patient information that doctors and nurses currently have to seek out or pull from a jumble of separate information sources will be automatically pushed to providers on personal digital assistants, smart phones or laptops in a predigested, visually oriented format. The most dramatic changes, however, will come in the way we treat diseases.

I recently took care of a 75-year-old gentleman who had been to see his internist about three months earlier for a routine visit and was subsequently found to have prostate cancer. Owen Hammond was a very healthy—what we doctors would call a *good*—75. One of the remarkable things that become apparent in a medical career is how different two people born in the same year can look after the passage of the better part of a lifetime. We see fifty-year-olds who look eighty, and eighty-year-olds who look fifty; and while some of the differences are genetic, most are the payoff of a lifetime of what might be called good health housekeeping. One can pretty well predict that the eighty-year-old who looks fifty hasn't smoked, has gotten a lot of exercise, doesn't have diabetes or hypertension, keeps mentally engaged and almost certainly hasn't spent a lot of time being a patient. In other words, that individual has kept the physical and mental house in order.

Hammond was a compact black man, five feet, six inches in height, who weighed about 150 pounds. He had an enduring smile and no frown lines. I never saw him in anything other than a suit and tie, with cuff links and a tie pin. My first impression was that he was one of those men, often small in stature, who seem to have more energy and enthusiasm than can comfortably fit inside their bodies. He took a forward-leaning approach to life.

While there is a certain caution to the way many people interact with doctors, every bit of Mr. Hammond was right there in our interaction. He gripped firmly, made eye contact and punctuated every sentence. He allowed as how he was active in his church, played tennis two or three days a week and walked a mile where others would have taken a cab. He'd never smoked except for a brief period of experimentation several decades earlier, but he had consumed *plenty* of good red wine and was still working every day, although most of his "work" in these later years was charitably oriented in conjunction with his church.

All in all, he had a flexible, optimistic attitude about life and life's wrinkles, so that when his screening prostate-specific antigen test or PSA, obtained as part of his routine physical examination, came back quite elevated, suggesting the possibility that he might have prostate cancer, he handled the whole situation with equanimity. He was friendly with several other men with prostate cancer, one of whom had died from it, but he knew enough to realize that the diagnosis wasn't a death sentence. As it turned out, a subsequent ultrasound showed that he had a prostate lump, and the biopsy was positive for cancer. Fortunately, the rest of the test showed that it hadn't spread.

Like many men of his age, Owen Hammond had already mentally prepared himself for *some* kind of bad health news and was, frankly, more than a little relieved to find out that it wasn't terminal. Prostate cancer is the most commonly diagnosed cancer in men in the United States, and the second most common cancer in men in the world, although the incidence in westernized countries is ten times that of countries that don't follow the western lifestyle. We don't know exactly why this is the case, but the presumption is that diet plays a major role. Prostate cancer rates are increasing in many countries, although most investigators believe this is due to better screening rather than real changes in disease incidence. Of course, it's also possible that the increase is real and due to increased exposure to whatever causes the disease in the first place. Survival rates are very good if the disease is diagnosed while it is still localized to the prostate; however, prostate cancer is still the second leading cause of cancer deaths among men in the world.

Prostate cancer is usually discovered either as a result of an elevated PSA blood test, which is one of the routine screening tests recommended by the American Cancer Society and many other European medical associations, or due to abnormal findings on physical examination of the prostate—abbreviated as DRE. The latter is considerably less pleasant for many men than the needle-stick required for the PSA, because a DRE is the scientific euphemism for digital rectal examination; and without going into unnecessary and probably unpalatable detail, a competent DRE involves a glove, the index finger and, ideally, a fair amount of lubricant jelly. The doctor performs a digital prostate examination as part of an adult male's routine visit. The exam is often the first way to catch prostate cancer, and is based on the identification of one or more nodules on the normally smooth, walnut-sized gland inside the rectum. If the PSA blood value is high or the rectal examination is abnormal, a prostate ultrasound is done, followed by a biopsy if appropriate.

If the biopsy shows cancer, there are a variety of treatments. When the cancer cells are limited to the prostate and haven't yet spread elsewhere, a patient can choose among several techniques for surgical removal: external beam or implanted radiation, cryosurgery with cold argon gas or treatment with highly focused ultrasound beams. For the philosophically inclined and for elderly patients with slow-growing, low-grade tumors, there is the Zen-like watchful waiting. If the disease has spread, there is chemotherapy or hormonal treatments. A recent report from the Agency for Healthcare Research and Quality concluded that no one prostate treatment is demonstrably superior to the others, and they all have pretty good results. As one of the report's lead authors, Dr. Timothy Wilt from the Minneapolis Veterans Affairs Center for Chronic Disease Outcomes Research, put it, "Having been involved in this area for a long time, it was not shocking, but it is disappointing . . . Information is really lacking to determine whether, over all, one treatment is more effective and preferred."

The expression "All roads lead to Rome" is obviously Rome-centric, and it turns out that Romans built a lot of roads going off in all different directions. The emperor Augustus constructed the golden milestone, or *milarium aurium* near the temple of Saturn, and declared that all distances would henceforth be measured from this bronze phallic symbol. Like the emperor, some physicians who treat prostate cancer have single-minded views about the best approach— often one that aligns with their economic interests—but most acknowledge that many reasonable alternatives exist, and many patients take one of the middle roads. Fortunately, as with Roman highways, all of the prostate treatment paths seem to end up at pretty much the same location.

Mr. Hammond's urologist was an unbiased advisor and ran through the list of the most common treatments with their potential risks and benefits. He included prostate removal, or prostatectomy, which can be done in a variety of ways; minimally invasive prostatectomy is increasingly preferred at advanced centers. He also mentioned several radiation alternatives, although proton beam treatment wasn't on his list because it wasn't (and still isn't) widely available. The disease was localized and hormonal therapy was therefore not appropriate.

They also discussed the reasonable alternative of doing nothing—which didn't appeal to either of them.

As he worked his way through his alternatives after talking to the doctor, Hammond had to weigh each option against some very personal issues that, while seemingly tangential, had a direct bearing on his final decision. While he loved his wife, sex no longer played a major role in their relationship because she too had health problems, so he wasn't undone by the potential risk of impotence. He wasn't afraid of surgery but he didn't really want to go through an operation because he and his wife had planned a cruise on the Mediterranean in a couple of months. He vowed to be as fit as possible for the trip. He was also a very well-groomed man and was viscerally put off by the treatments that might leave him with a tendency to dribble urine into his underwear every time he coughed or sneezed—commonly known as stress incontinence. In the end, following a fair amount of discussion with his wife and some of his male friends, and after doing some of his own research on the Internet, he opted for a relatively conservative approach, which was the implantation of a hundred or so radioactive seeds throughout his prostate.

The seeds are placed with a needle through the skin behind the scrotum. They contain radioactive iodine and give off low-dose x-rays that pretty much stay within the prostate. There is no risk to others and the radiation eventually wears off as the x-rays in the radioactive elements become depleted. Owen Hammond had the seeds placed with spinal anesthesia as an outpatient; and with the exception of some short-lived diarrhea and incontinence, he had limited side effects. Had proton beam therapy been available at the time he was making his decision, he would in all likelihood have opted for that quick, painless, high-tech treatment. As it turned out, he and his wife went on their cruise without problems, and after a year, he had pretty much stopped thinking about the cancer as an issue.

The variety of treatments for prostate cancer that we can offer patients like Mr. Hammond might be described as an embarrassment of riches, but as with many industries, medicine is moving away from an off-the-rack approach to one much more like that of the Savile Row bespoke suit. The era of *personalized* medicine has arrived, in which big medical centers behave like large corporations in their competition for consumer-patients. Large medical centers will try to offer as many different, cutting-edge treatment alternatives as possible because if you have a choice between going to a hospital that can offer you both a robotic prostatectomy and proton beam therapy or going to a hospital that doesn't have either, you're probably going to go to the one with more options. Today's hospitals are in the midst of a period of rapid change that began within the past three decades. A hospital thirty years ago could afford to act like a lumbering herbivore; the hospital of the future will be a very different animal—efficient, attractive, omnivorous and perhaps even predatory.

Decision support systems, data visualization tools, intelligent software and modeling tools are beginning to take on a growing role in the diagnosis and treatment of patients as well as in the business of medicine. Let's imagine, for example, a hypothetical asthmatic nineteen-year-old college student named Callaghan who's off to the beach with a car full of friends for a week in the sun. Let's say the driver has a cat, unbeknown to Cal. The driver, of course, doesn't know about Cal's asthma and that he's extremely allergic to cat dander. About four hours into the drive, Cal begins to feel tightness in his chest. His inhaler doesn't help, and by the time everyone else in the car is persuaded that they need to break off the trip to take him to the hospital—to a place most of them have never been before—he's in a little bit of trouble. He can't talk when he gets to the receiving ward, and none of his friends can help with his medical history, so the doctors have to start from scratch, treating him with the how-I-do-it approach they've always used. However, the first medicine they use to treat the asthma causes Cal's heart to take off like a racehorse, just as it did the last three times he got it unbeknown to the doctors. Eventually things get so bad that he needs to be put on a breathing machine. Fortunately, after a couple of days of how-I-do-it management, things turn around, and the doctors are able to decrease the amount of breathing support given by the respirator. Eventually, Cal catches up with his buddies at the beach a week late.

When Cal arrives at the hospital of the future, however, things will be very different. Using some identifier, perhaps a radio-frequency identification tag implanted under Cal's skin, any doctor at any hospital in the world will have access to his medical history and allergies. They'll know immediately that he shouldn't get the drug that he previously reacted poorly to. The hospital's computerized patient management software will have an embedded decision support system that will automatically start an up-to-date, consensus best-practice treatment designed by the world's experts on the disease, and provide the ability to fine-tune it based on such things as Cal's size and weight. In essence, Cal's doctors will be able to manage many aspects of his care in the same way that modern commercial aviators fly their vehicles on autopilot. To be sure, they'll be there to step in when needed, just as pilots do, but the system will keep track of Cal's care automatically, consistently, and will free the doctors to be more efficient, perhaps caring for other sick patients.

Similarly, laboratory results, breathing tests, x-rays and physiological data will be molded into a coherent, readily accessible, three-dimensional picture, allowing the physicians to monitor Cal's current state at a glance without needing to comb through all of the data piecemeal. They'll be able to see this data from wherever they are on handheld display screens. Problematic trends will be automatically brought to the attention of the doctors by smart software. And if Cal still needs to be put on a breathing machine, a smart respirator will manage its own settings rather than awaiting interventions from busy doctors, respiratory therapists and nurses. When he's strong enough to breathe on his own, the ventilator will automatically summon someone to come and remove the breathing tube. Finally, doctors will have access to computer-based decision modeling tools to virtually explore potential treatment alternatives before actually trying them on Cal.

Medicine—in addition to increased standardization, computerization and robotics—like many other industries, is poised to become global. Medical tourism is on the rise: Today patients travel to resort locations for plastic surgery "vacations" in the sun and come back remade; companies send employees offshore for joint replacement and heart surgery at steep discounts; and desperate parents fly across the world with their children to get stem cell treatments unavailable in their own countries. As performance and pricing information become globally available on the Internet, successful health systems of the future will increasingly market to both national and international clientele, perhaps by the creation of satellite locations providing their uniform, brand-certified level of care, or perhaps by creating incentives for travel to their core facilities. The Cleveland Clinic, for example, has marketed its name nationally and attracts a wealthy international clientele to Cleveland, where they can stay in a first-class hotel on the hospital grounds. The Mayo Clinic has opened satellite hospitals in sun belt locations in Scottsdale, Arizona, and Jacksonville, Florida. As cost and performance information become more transparent and publicly accessible, centers of excellence for the management of certain conditions will emerge, and the customers—be they employers, insurers or patients—will gravitate toward them.

However, the critical index of a hospital's health will remain its financial integrity. The *sine qua non* for the hospital of the future will be its ability to invest in the new transformative technologies of computer systems, robotics, communications, quality control and tracking systems. Such innovations will allow hospitals to stay competitive even as financial water holes inevitably dry up and ballooning medical costs around the globe force us to focus our efforts on cost-effective, efficacious treatments.

As we'll see in the rest of this book, we are in the midst of an evolutionary explosion in medical technology. Diversification is occurring despite—or perhaps as a result of—growing resource constraints. We may later compare this rapidly changing medical era to analogous periods in true evolution that appear to have been precipitated by dramatic changes in the environment. An optimist's view, and one that I hope you'll share by the end of this book, is that the medicine of the future will be preventative rather than the rear-guard action it too often seems to be today; that it will therefore be less expensive; that the hospitals of the future will be necessary only for the very ill; that there won't be nearly as many of them. And, that, when we eventually emerge from the fog, we'll be able to do a lot more, for a lot more of us, for a lot less.

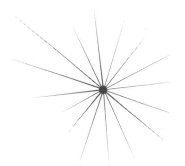

CHAPTER 2

EYE IN THE SKY

At about three o'clock on a recent morning, I was doing my rounds on seventy-odd patients in five intensive care units (ICUs) scattered around Philadelphia . . . all from an office building several blocks from any of them. The room in which I was working looks a little like a military bunker, with several people seated in front of banks of computer screens, but there is also a tank full of brightly colored tropical fish in one corner and several bulletin boards scattered around the room are full of thumbtacked medical articles. I was acting in the role of what we euphemistically call a "doc-in-the-box"—sitting at a workstation monitoring five flat-panel displays with patient information, patient alarms and audio-video links to the patients themselves. I was running a telemedical intensive care service, a new technology that allows a small team of nurses and doctors to simultaneously care for more than a hundred patients in geographically dispersed hospitals.

Here's the way a telemedical ICU works. All of the patient's vital signs and lab studies, such as the heart rate, blood pressure, temperature, breathing rate, blood counts, chemistries, oxygen and carbon dioxide levels, are displayed on a workstation. So are the nurses' and doctors' notes from the bedsides. All sorts of other important information, like electrocardiograms, chest x-rays and CAT scans can be pulled up on a monitor to be evaluated by the doc-in-the-box. Additionally, each patient's room has a microphone and camera mounted on the wall above and opposite the patient with which the doctor can look at the patient, and talk to him or to the doctors and nurses in the room. The cameras can be steered to look around the room and zoom in to evaluate small details, such as the tubing in a patient's nose or mouth.

Our doc-in-the-box sits at his workstation next to an intensive care nurse and the two engage in continuous rounds all night long, visiting patients via the cameras, zooming in on trouble spots and working in coordination with the various bedside teams, which typically consist of a doctor in training and a nurse. At the University of Pennsylvania, the telemedical ICU covers the hours between seven at night and seven in the morning with only faculty intensivists (doctors who specialize in the care of critically ill patients in ICUs), although our expert ICU nurses are there most of the day as well as during the night. Other telemedical ICUs cover more beds and run round the clock; some cover many ICUs in several states from the same central location.

At three o'clock in the morning most senior physicians are at home in bed, with the obvious exception of a few in locations such as the emergency room, operating room and delivery room. At 3 AM, whether in a small community hospital or a large academic medical center, the sickest patients in the intensive care unit are typically managed by house officers or nurses. They operate according to standing protocols or phoned-in orders, often from a sleepy senior doctor who has just been woken up and given a brief summary of a problem. Perhaps it's because of the darkness, perhaps because the usual bustle of the ICU is absent and probably because of the life and death nature of the events that often seem to happen then, but night time in the ICU has always put me in mind of the old prayer: "From ghoulies and ghosties and long-leggity beasties, and things that go bump in the night, good Lord deliver us."

On this particular evening, a smart alarm went off on one of the many patients I was covering: Beth Dougherty had just had a heart valve replaced several hours earlier, in the afternoon of what was now the day before. Dougherty was pretty young for this operation—she was 55—but had some special circumstances. She had been born with a congenitally abnormal heart, and she'd already been through heart surgery when she was a child. For reasons I'll describe later in the book, artificial heart valves have a limited life-span. Many people who undergo heart surgery for a congenital problem will need a second surgery, and perhaps even more, later on. And unlike a lot of things in life, heart surgery doesn't get easier the second time around.

Dougherty wasn't enthusiastic about having another major operation. Like many people who have gone through major illness and surgery as a child, she hadn't really had a normal childhood and was considered delicate by her family and friends. She didn't like going to the hospital to see doctors, and she, more so than most patients, had good reason. She'd always known that at some point her valve would need to be replaced—the old valve was of a dated design, and she could actually hear it ticking quite audibly when she lay in bed each night. While she'd never been terribly athletic, she had recently found that she couldn't manage a flight of stairs without stopping, nor could she carry the groceries into the house.

An echocardiogram done by her cardiologist showed that a shelf of thick, calcified tissue had formed under the old valve. This growth is called a pannus and results from a chronic inflammatory reaction to the artificial material in the valve. The shelf had begun to block blood flowing from her heart and there was no question that she needed a new valve—or one day soon she would die, very possibly suddenly.

On the day of her surgery, Beth Dougherty had been rolled into the operating room on her stretcher, looking fearful but obviously maintaining a sense of humor. She told the OR nurses she was an old pro at this; and she wore what's known as a scapular—two laminated pictures of saints attached to one another by lengths of brown or green cloth. She asked the nurses to make sure that she had contact with her brown scapular during her operation, and the nurses tied it loosely around an ankle. Pope John Paul II had done the same during surgery following his attempted assassination in 1981.

Surgery of any kind leaves scars—on the outside *and* the inside. The way the body repairs itself is by the formation of a clot, which is eventually replaced by collagen, or scar tissue. After you have one operation, a second operation in the same location is much more challenging for the surgeons because all of the usual anatomic "landmarks" are covered in scar tissue. This scar makes it very difficult for a surgeon to find her way around and occasionally causes major missteps. For example, where a major nerve would ordinarily be very obvious to the surgeon on a first operation, that same nerve may look exactly like the scar tissue surrounding it when the area is revisited. A big problem for heart surgeons operating on an area for the second time around is that something looking like a scar may actually be the wall of the heart, and the surgeon may inadvertently cut into a heart chamber, which is often followed by a geyser of blood. Another problem with a scar is that it is supplied by lots of tiny little blood vessels, which all bleed when cut.

When we prepare to operate on a patient's heart for the first time, we go through a regimented process to minimize the possibility of infection and to set up the incision. Electrocardiogram leads are placed at five places on the torso; for a male, the chest is shaved. Then we use two different liquid preparation solutions to kill the bacteria on the skin. Once the chest is dry, a sheet of adhesive plastic impregnated with an iodine-like chemical is placed over the breastbone, or sternum, sticky side down. Finally we lay a set of azure-colored, Gore-Tex drapes around the chest, clipping them to one another with a type of stainless steel surgical forceps called an Allis. When all the prepping and draping is done, the rectangular area of prepared skin has a bronze shade from the prep and adhesive, and is surrounded by a blue from exactly the opposite side of the color wheel originally designed by Isaac Newton and therefore very pleasing to the eye. We complement these two colors with a third, a red pinstripe, which is the blood from the incision down the center of the chest. The cut is made with a scalpel, and an electronic cautery used to stop any bleeding causes a smoky haze to briefly drift over the operating table. A couple of more passes with the scalpel and the pearly white bone of the sternum is visible. At this point, the surgeon slips the tip of what looks like a modified band saw under one end of the sternum and cuts it in half with one steady, buzzing pull.

When the sternum is opened, it's often possible to see the heart beating away in its special cocoon, the pericardial sac. With the snip of a pair of scissors, the pericardium is opened and there lies the heart, looking nothing at all like the one tattooed on many patients, but beautiful nonetheless. I still feel a little surprised every time we reveal this animated organ inside a motionless human being.

With the exception of a little bleeding from the skin and the raw edges of the breastbone, this is usually a remarkably bloodless portion of the operation when we're operating on what we call a "virgin" chest. Things are quite a bit different when we go in for the second or third time around, as with Beth

Dougherty. First of all, we prepare differently. We put in more intravenous lines (IVs) so we can give blood quickly if things go wrong. We bring blood into the operating room and have it standing by on ice, in coolers, so we can start transfusing immediately if we cut into the heart or a major blood vessel. We put on special defibrillator pads as well, so that we can shock the heart if it starts fibrillating. And the surgeon approaches the heart with an extra amount of caution.

The saw we use for *re*operations is more like an archeological tool than a band saw; for example, the blade's edge is curved like the arc of a radial saw, but rather than spinning, it reverses direction rapidly back and forth, allowing the surgeon very precise control as he saws through the bone. Before he can start sawing, however, the wires that were used to hold the closed sternum together at the end of the last operation must be removed with pliers. Once the bone is completely cut, the two halves of the sternum are very carefully spread apart to avoid tearing open the heart, which lies just beneath, encased in scar tissue. Finally, the scar tissue is meticulously dissected away from the heart, eventually revealing the critical anatomic structures that are usually apparent as soon as the pericardial sac is open. This extra dissection may take an hour or more; and before the main operation even begins, the heart of a patient undergoing "redo" surgery is surrounded by cut and bleeding scar tissue, which looks more like raw meat than anything recognizable from an anatomy book. All of this tissue will continue to bleed at the end of the operation, complicating the postoperative management of the patient.

Using this careful approach, the surgeon assiduously opened Beth Dougherty's chest, and as the dissection proceeded, the clicking sound of the old valve got louder. He then replaced it with a newer valve and made sure she wasn't bleeding excessively when he reclosed the sternum. But, as with most patients, she continued to bleed into the drainage tubes we usually put around the heart after this operation.

When I made my virtual rounds that night in my role as the doc-in-the-box, several hours after Beth Dougherty's surgery I was moderately concerned about the amount of blood coming from the drainage tubes. As I investigated, I became even more concerned that her red blood count was far lower than could be accounted for by the amount of bleeding I could see in the tubes. Something out of the ordinary was clearly going on because the bleeding from around the heart normally slows to a stop over the first couple of hours after heart surgery. Not only was she bleeding a lot, but several of the numbers didn't add up right. Her blood hemoglobin was very low, even lower than it should have been from the blood collected in the drainage tubes; and the discrepancy suggested that more blood was going someplace I *couldn't* see. This required urgent investigation. Two typical places that can hide a lot of blood in the human body are the belly and the area around the lungs. I ordered a chest x-ray, which showed what looked like a collection of blood in the right side of the chest, in an area not drained by the tubes placed during her surgery.

I talked this over with the resident house officer who was on-call in that unit and told him to insert another drainage tube in the right chest. He did, and as I watched through the camera, the tube drained two and a half liters of blood, or the better part of a gallon, which represents about half of a normal person's *total* blood volume and fully accounted for the missing blood. It was immediately clear that Beth Dougherty needed to go back to the operating room immediately to be reexplored and have the source of the bleeding located. While this repair is often as simple as putting an extra stitch on a blood vessel, it sometimes means that something catastrophic has happened, and the whole operation has to be redone.

The advantage a doc-in-the-box has is a birds-eye view of the situation. A friend and colleague of mine, Ben Kohl, tells another story—one that might be amusing were it not for the fact that it illustrates a deadly serious near-miss—about an experience he had one night when he was the doc-in-the-box and how his ability to watch things from the virtual sky allowed him to help save a life. The nurse he was working with told him there was an alarm from the room in which a bedside team was in the midst of a procedure in one of the units. The procedure involved inserting a large intravenous catheter into a vein in a patient's chest, which is done with a full set of sterile surgical drapes covering the patient's head and torso to keep the catheter from being contaminated. Ben said he looked into the room with the camera and saw a couple of doctors in sterile gowns, gloves and masks beginning to anesthetize the skin over the vessel. The nurse was standing by and helping gather needed equipment. While Ben had intended just to glance in, some instinct bred by years of experience told him that something was wrong in the room. He paused and took a closer look, zooming in on the drapes over the patient's chest.

It is not uncommon to give a patient who is about to undergo a procedure like this a little sedation to make the whole process more bearable, and we often give an intravenous drug in the same family as valium or a narcotic. While these drugs can dull the experience of an unpleasant ICU procedure, they can also make a patient sleepy and may slow or stop breathing. When Ben looked at this patient's chest, he had subliminally noticed that it wasn't rising and falling; the patient had stopped breathing, so he asked the preoccupied and now startled team to stop the procedure immediately and undrape the patient. When they did, it became apparent that they needed to call for a "code blue" team to insert a breathing tube into the patient's lungs. The doctors and nurse at the bedside had been so focused on what they were doing, so heads-down, that they had missed the fact that their patient was getting into big trouble right under their eyes; and yet Ben was able to spot this from his eye in the sky in an office building miles away.

This ability to be one step back from the crowd—to see things from the middle distance rather than being sucked into the chaos in a code blue or missing the rapidity with which a problem is evolving—is one of the strengths of telemedical care. In the same way that Ben knew there was something wrong with the patient who had stopped breathing, I knew that Beth Dougherty had something bleeding around her heart that required immediate attention. After having dealt with some other issues that night following Dougherty's surgery, I looped back to make sure that the resident had actually contacted the surgeon about urgently reexploring her wound to find the source of the bleeding. The resident said—a little evasively, I thought—that he had indeed gotten in touch, but when I pressed him a bit, it turned out that his method of getting in touch about this emergency situation was to text message the surgeon. Now I happen to know this surgeon pretty well, assumed he was asleep rather than sitting bleary-eyed staring at his cellphone, and so I *called* him, told him what was up, and we agreed that she needed to be taken back to the operating room immediately. Not surprisingly, he hadn't yet seen the text message.

As she was being wheeled down the hallway to the operating room, the blood abruptly stopped draining, which might sound like a good thing; but it actually meant that the tubes had clotted off. Dougherty's blood pressure had become barely measurable and they hustled her onto the OR table. The undrained blood was now building up pressure around her heart like a tourniquet rather than emptying into the tubes; and it was only by virtue of some extremely fast action on the part of the surgical team that they were able to open her chest rapidly and scoop the clot from around the compressed heart, saving her life. The bleeding source was eventually found around the sewing ring of the new heart valve. It was fixed without much difficulty and the rest of her hospital stay went smoothly. Her breathing tube was removed a few hours after she returned from her second trip to the operating room, and she left the hospital pretty much on schedule. She never realized quite how close she had come to dying.

These scenarios perfectly demonstrate some of the promises and pitfalls of computer networks in the delivery of medical care. The evolution of networks, such as our hospital intranet, permitted the development of a telemedical ICU; and teleintensive care is but one form of telemedicine, the promise-filled new medical technology that can extend the geographic and temporal reach of expert physicians, be they pathologists, radiologists, intensivists, surgeons, pediatricians or internists.

Assisted by smart software, I recognized that Beth Dougherty was bleeding that night, which led to the insertion of a chest tube and, ultimately, to the patient's timely return to the OR. Before the advent of networks, the bleeding problem might have gone unrecognized until something catastrophic happened; on the other hand, before networks, the resident would have been forced to make an actual phone call rather than feeling he'd discharged his duty by text messaging his attending doctor. In many ways, networks in medicine are like automated voice-answering systems that can, if designed with good intentions, funnel you very efficiently to the right person to solve your problem or, if implemented poorly, can dump your message into some anonymous, voice-mail dead-drop.

While one would think that telemedicine was a brand new technology, it turns out that telecardiology and telestethoscopy date back a hundred years to radio- and telephone-based applications at the turn of the last century. Early pioneers transmitted the sounds of the human heart over the radio and through the phone, although almost certainly without enough fidelity to make them useful. There was even a demonstration of telepsychiatry in the 1950s, in which psychiatric services were provided to remote rural areas. In fact, telepsychiatry is still used today to serve rural areas and prisons, and now employs video-conferencing technology.

The most successful and widely used example of a "store-and-forward" telemedical application capitalizes on the half a day's time difference between Australia and much of the United States. Store-and-forward systems acquire and store data, such as x-ray images, on site at a hospital and forward a copy to a remote offsite location for analysis. The store-and-forward applications aren't so critical to patient care that the person at the far end of the link needs to be in immediate, live linkage with the local site; but they are much better than old-fashioned approaches in which there might be many hours between the performance of a test and its analysis.

Radiology is an example of a medical field ideally adapted to store-and-forward telemedicine. Dr. Paul Berger was the former head of an expert-witness consulting firm, and then of a large radiology group practice in California that served a number of hospitals. Many of Berger's clients required twenty-four-hour expert radiology coverage, because in today's world it won't do to have a radiologist tell you a patient has a blood clot around the brain in the morning, when the CT scan was actually performed by a technician the night before. Berger's first cut at a solution involved the development of a single, central x-ray reading site located in California. After a time it became apparent that it was too costly to hire and retain radiologists willing to work nights; they either grew tired of the night work or wanted too much money.

During this period, Berger engaged the services of a world-renowned magnetic resonance imaging expert, Dr. William Bradley. At one point his radiographic interpretation expertise was required in the middle of a California night, when Bradley happened to be lecturing in the middle of the day in China, half a world away. As with many pioneering medical innovations, creative people find ways when someone's life is in the balance. The images were sent over the Internet to China, where Bradley was able to provide an immediate reading. The light went on for Paul Berger and a company named

Nighthawk Radiology Services was born. Berger set up a radiology reading center in Sydney, Australia, and staffed it with U.S.-credentialed radiologists who, in many cases, relocated there from the States. Today, Nighthawk provides nighttime radiographic interpretations for almost 1,500 American hospitals, using radiologists working daylight shifts in Sydney, and another reading hub has been opened in Switzerland.

Radiologists who practice teleradiology have no problems getting credentialed to practice telemedicine at each of the hospitals they cover because it is in the hospital's interest to have that coverage. A telemedical radiologist may be licensed to practice in as many as forty or fifty states; and malpractice coverage is merely a matter of finding an insurance carrier that can figure out an actuarial dollar value that works for all of the covered locations. Consequently, a credentialed, licensed, board-certified radiologist with malpractice coverage can cover any hospital in the United States from any location in the world, provided he has access to the right viewing equipment, which amounts to a high-resolution computer monitor, the appropriate software to manipulate the radiographic images, and a reasonably fast Internet connection. Images are gathered from the local x-ray machines, sent to the teleradiology company's server for *storage,* and then *forwarded* to the radiologist wherever in the world she happens to be. In a very real sense, teleradiology is in the vanguard of the globalization of medicine.

The store-and-forward model works perfectly for image-based telemedical services such as radiology and pathology, for which there is no need for a live radiologist or pathologist to be at the other end of a real-time connection. As long as the remote radiologist performs a reading within a reasonable, perhaps contractually stipulated, time period, patient care is well served. In contrast, other telemedicine applications, such as intensive care medicine, require immediate access to a live body: It is of no benefit to send an alert from an ICU to a remote intensive care specialist if he can't get back to you for an hour; and it turns out there's a lot of data showing that immediate access to an intensivist makes a big difference in patient outcome.

Two intensivists who used to work in a medical center, Dr. Mike Breslow and Dr. Brian Rosenfeld, recognized the potential benefits of telemedical intensive care ten years ago when they were on the faculty at Johns Hopkins

University Medical School in the Department of Anesthesiology and Critical Care Medicine. Breslow is a very focused, forward-thinking man who graduated from Harvard and Tufts Medical School and went on to become board certified in three different fields—internal medicine, anesthesiology and critical care medicine. He *looks* studious, and he tends to pause before he answers a question, taking time to mull over the answer carefully. Rosenfeld is superficially more of a hail-fellow-well-met kind of guy, but the appearance is deceptive. He trained in internal medicine, pulmonary medicine, anesthesiology and critical care. Both of these doctors are extraordinarily well trained and both had been practicing academic intensivists for two decades when they embarked on what seemed at the time to be a routine academic study. They performed a four-month evaluation at a Hopkins-affiliated hospital where they implemented round-the-clock coverage of an ICU using remote telemedical intensive care.

The study showed pretty dramatic reductions in patient complications, costs, length of stay in the ICU, and mortality. As with Berger and Nighthawk, the light went on for Breslow and Rosenfeld. Curiously, when they brought this concept to the Hopkins technology transfer office to give the university a chance to commercialize the idea, which they were contractually obligated to do as faculty members, they were told, in effect, that it would never fly. In a fairly daring leap of faith, they left Hopkins and the safe confines of one of the best academic medical centers in the world to start a new company they called ICUSA. Their company was recently acquired for several hundred-million dollars by Philips Healthcare.

ICU telemedicine makes good sense both for improved patient outcomes and for a pressing problem. A large recent study jointly performed by all the major professional societies with an interest in critical care showed that we are headed into an ICU-manpower crisis. We know that intensivists improve outcomes and reduce costs. We know that the demands for intensive care services will increase as the baby-boomer population ages. We know that intensive care is expensive—some estimates suggest that as much as ten percent of a hospital's budget is spent on intensive care. And we aren't training enough intensivists to fill the projected demand. Intensive care telemedicine is therefore an ideal solution for a number of reasons.

A typical intensive care unit in a medium-sized hospital has ten or fifteen beds, cares for both medical and surgical patients, and is the only such unit in the hospital. Larger academic hospitals may have many ICUs, but community hospitals typically have just one. Most of the 55,000 ICU beds in the United States are in hospitals with between 100 and 300 total beds and are not covered by an intensive care specialist at all; rather, the patients in these units are cared for by committees of doctors who have busy outpatient practices and who stop by the ICU only once or twice a day to make recommendations about the organ system in which they specialize—and the sicker the patient, the larger the committee. Larger hospitals, like mine, have several specialty ICUs, such as cardiac surgical, trauma, neurosurgical, medical and neonatal units, with trained intensive care teams in each. However, even in these so-called high intensity units, a bedside intensivist can effectively cover only about 15 patients at a time. More than that and the combined burdens of data analysis, documentation, medical procedures and crisis management become overwhelming.

Intensive care telemedicine has something to offer to both the community hospital with one ICU and the large academic medical center with several specialty ICUs. The medium-sized hospital may implement a telemedical ICU program to ensure that an intensive care team is looking after its 10 or 15 sickest patients because it's not cost-effective to hire an on-site intensivist to work fulltime in a single, low-acuity unit. Large academic hospitals, on the other hand, have units that are packed with very sick patients who require constant, expert round-the-clock attention. At our hospital, for example, patients are admitted to the ICUs all night long, from the emergency room, the trauma bay, the operating room and the helipad. A single telemedical intensivist, working with experienced critical care nurses, can effectively cover over 100 ICU beds simultaneously—and those beds may be scattered among dozens of hospitals, of both the community and academic type.

I've given you a couple examples of how an off-site intensivist can intervene to avoid catastrophe in the ICU, but we use telemedicine for a wide variety of seemingly more mundane, yet equally important ways. We are an academic institution and it's not uncommon for us to get a phone call from a trainee who wants to run a question by us for help. Nurses may call to say that they're off to take a patient somewhere else in the hospital for a study, and can we look more closely after another patient they're assigned to while they're gone. Occasionally patients are agitated and pull at tubes or try to get out of bed, and we turn the camera on in that room to watch the patient continuously so we can call someone into the room before they get into trouble. We've lowered lung infection rates in one of our ICUs dramatically by running through a checklist a couple of times a day to make sure all of the appropriate preventative measures are being taken, such as ensuring that the head of the bed is tilted above thirty degrees, which prevents stomach contents from getting into the lungs.

One obvious question is why a telemedical intensivist can cover many more patients than the intensivist at the bedside. The best way to understand how a doctor sitting at a computer workstation in an office building can be more efficient than someone working in a physical ICU is to compare the workflows of each. When I first started practicing intensive care medicine, we began rounds by filling up a rolling, slotted cart with patient charts, which were merely looseleaf binders packed with information and organized under tabbed dividers. Each ICU patient had a regular chart and a second binder dedicated to vital signs, such as blood pressure and heart rate, of which there are many in the ICU. As we proceeded from one bedside to the next, we'd pull along the cart of charts and open the binders, laboriously leafing our way through all of the data acquired over the previous twenty-four hours. Sometimes critical pieces of paper had torn out of the chart and couldn't be found. Sometimes another doctor needed the same binder we were reading, so one of us waited for the other. Sometimes, as is the way with binders, they dropped, and all of the papers fell out, which made for a bad morning.

The vital signs were charted on fold-out paper, like centerfolds, only the information contained therein was much less eye-catching than their glossy magazine counterparts. Even today, in most ICUs, nurses transcribe data from sophisticated computerized monitors onto similar foldout records by hand, which I can only liken to some Victorian clerk furiously trying to scribble down stock data with his quill pen on the floor of one of today's high-tech stock exchanges. When we finished interpreting and analyzing the data on a given patient, we would write an often illegible note, enter some equally illegible orders and scurry to the next bedside.

Unlike the way I used to do rounds, as the doc-in-the-box, telemedical intensivist, I sit in front of a tidy array of computer screens, hyperlinking and multitasking my way through virtual patient charts with single clicks of a mouse. Want the results of a lab test? Click. What's the hemoglobin? Click. Want to see the trend of the same lab test over the past week? Click. Want to actually see the *patient?* Click. Want to write an order? Click again. This is obviously a much more efficient way to check on patients than the old-fashioned way: My notes are legible, *and* several doctors can look at the same chart at the same time. While I'm making rounds, smart software is working away in the background, culling through the data streams looking for anomalies or worrisome trends. This smart software continuously feeds patient alerts from all of the patients I'm watching onto a virtual conveyor belt scrolling across one of the screens: "The patient in room 7 at Memorial Hospital has a low blood pressure" and "The patient in room 15 at Community General is breathing fast" and "The patient in the recovery room in bed 4 has a falling hemoglobin."

The doc-in-the-box is often the first to know about evolving patient problems because she's effectively in the crow's nest with a higher dimensional view of the landscape. For her, the world is not flat.

A friend of mine recently told me he'd got a cold call from someone who wanted to buy his house—which was surprising because he hadn't put it on the market. The caller was a very wealthy man from out-of-town who had evidently hopped in a helicopter with his realtor and flown over the area to which he planned to move, looking for interesting properties. The rich topographer spotted my friend's house from the sky, did a little research and made him an offer he couldn't refuse.

That guy in the helicopter is a lot like the telemedical intensivist with the smart software and the camera: Both have access to highly efficient search and discovery engines and can very quickly find the information they're looking for using their eyes in the sky. We can't completely do away with the services of the bedside doctor—intensive care patients need bedside doctors to examine them, do procedures and meet face-to-face with families, but complementary teams of bedside and telemedical specialists provide the advantage of round-the-clock coverage by experts.

A variety of other equally exciting applications of networks in medical care are under development. On September 7, 2001, Dr. Jacques Marescaux, operating from New York, successfully removed the gall bladder of a woman on the other side of the Atlantic, in Strasbourg, France, using a remote-controlled robot-assisted laparoscopic device. Since then, surgeons have performed more complex operations on remote patients using a variety of surgical tools. There have even been demonstrations of the feasibility of extraterrestrial telesurgery.

NASA's Extreme Environment Mission Operation (NEEMO) is based on the ocean floor off Key Largo in Florida. The base unit, called Aquarius, is seventy feet underwater; rotating crews of aquanauts live aboard for extended periods in conditions that simulate outer space. On one recent mission, an earthbound surgeon, using a surgical robot and located in Hamilton, Ontario, sutured a laceration on one of the Aquarius's underwater aquanauts. The device the surgeon controlled was a small portable robot, equipped with a camera and pincers, that would fit readily aboard a space shuttle. The operation was designed to determine whether an earthbound surgeon could operate on an astronaut located, for example, on the International Space Station.

The surgeon was Dr. Mehran Anvari, who directs the Centre for Minimal Access Surgery at McMaster University and whose experience in doing both remote laparoscopic and robotic surgeries on patients in rural Canada uniquely qualified him for the task. From a console at the university, he operates on patients hundreds of miles away and, in the case of NEEMO, almost a hundred feet below the earth's surface. We'll talk more about robotic surgery in a later chapter, but the potential synergies of networks and robots in medicine are enormous.

While teleradiology, telemedical intensive care and telesurgery are all pretty eye-catching examples of what networks can do in medicine, there are more things under the hood. Consider some of the network-based capabilities that we're beginning to see in new high-end cars. A variety of sensors acquire information about the engine, tires and road conditions. Braking and wheel speed are computer-controlled to prevent skids and spin. Onboard computers store information about speed, braking and steering in the seconds prior to an accident for later reconstruction. Many cars are equipped to wirelessly communicate with your cellphone, and some come with voice-controlled music players or global positioning systems that enable you to find your way. Radio-frequency identification tags (RFID) are built into on-board toll-paying sensors.

Network-based technologies in cars give us an idea of what's in store for medicine in the very near future. In our operating rooms at the University of Pennsylvania, for example, some of the major pieces of equipment are tagged with RFID devices linked to a virtual Web-based map of the operating room. If we happen to misplace an expensive piece of equipment, like a neodymium-doped yttrium aluminum garnet laser, which we affectionately call the Nd:YAG laser, we go to the map and find out where it's hiding. We use this same technology to track operating room visitors, and we insist that visiting salesmen and doctors wear RFID-tagged fluorescent vests, so that if someone wants to find the prosthesis salesman who was just lurking around where he shouldn't have been a couple of minutes ago, we can track him down. Many other devices in operating rooms and ICUs communicate wirelessly with the networks in the hospital, and voice-controlled surgical devices are already in use in some ORs. In the future, anesthesiologists may actually direct care in several operating rooms from networked anesthesia control centers. Similarly, surgeons may one day operate from surgical control centers controlling remote robotic instruments.

This is all great, but soon we may also have patient room sensors that continuously sample the air for the smell of infection, similar to the devices that sniff luggage for explosives at the airport. Secure access to hospital networks will be strictly controlled using biometric identifiers, such as fingerprints or retinal scans. Bedside monitors and devices will automatically and wirelessly "dump" their data into the laptop or personal digital assistant of an authorized physician or nurse as he passes by on rounds. And, as in the new automotive sensor-controller networks, the medical monitors will eventually be networked together with infusion pumps, mechanical ventilators and anesthesia machines so that they can prevent the medical equivalent of vehicular skidding or spinning out. In the next chapter, we'll see that the same sorts of smart, safety-enhancing computer algorithms are equally applicable to automobile and medical components. Advanced safety-oriented medical applications like these are well within the technological capabilities of today's computing and sensing devices, and they are likely to evolve rapidly as they become interconnected.

Biologists once claimed that ontogeny recapitulates phylogeny, which means that the growth of a human being from an embryo to an adult mimics our species' evolution from a single-celled organism to a complex system. Similarly, the evolution of the Internet we know today in some ways mimics the evolution of the mammalian nervous system. The primitive dorsal nerve cord was a simple nervous system, and the ancestor of the spinal cord that ran from one end of early organisms to the other. The dorsal nerve is analogous to the original, what we would now call primitive, backbone computer network that ran among research universities in the 1970s. Similarly, the local office networks of the 1980s linking word processors with printers were like the isolated cells and organelles of primitive organisms. At some pivotal moment in the evolution of the organism, Mother Nature discovered that good things happen when cells are able to communicate rapidly with other cells in the body, over long distances, through that primitive dorsal nerve cord. In the same way, at a similarly pivotal point, we learned that it was good for business if word processors could be made to communicate with other faraway word processors over a core, backbone network . . . and the Internet was born.

This is where we are headed in medicine as we proceed from a localized and provincial patient care model to one in which backbone medical networks ramify toward bedside medical devices, eventually forming a continuous web in which any authorized human or computer node on the network has ready access to required medical information from any other node.

There is an obvious tension between ensuring that medical information and skills flow freely across networks and ensuring that medical information remains private and secure. However, the same thing can be said for industries like banking, which have already computerized securely. The development of the universal medical record, such as the versions offered by Google and Microsoft, and of other currently stand-alone, medically oriented backbones, such as pharmacy networks, will enable evolution in the same way that the ATM networks did in the 1980s. An ATM in one part of a country might not have talked to a bank several states away twenty years ago; but it's now possible to find an ATM almost anywhere in the world from which to draw money from any bank. For the same sorts of reasons that it makes sense to have ready personal access to money wherever you are, it also makes sense to have immediate access to important medical information when needed.

Similarly, for the same reason that people seek out familiar restaurant brands like Starbucks or McDonalds in a strange country (under the assumption that the food safety and quality meets a certain standard), we will eventually find brand-name medical clinics in major cities all over the world, where a patient can count on the quality of the medications and the integrity of the care. And the care at those clinics will almost certainly be integrated in some part using networks linked to central "headquarters" maintained by the company owning that brand. The remote medical clinic doctor, for example, may consult with an expert specialist back at headquarters over the medical network. Your medical records, vital signs and laboratory results may be analyzed or tracked in computers maintained at headquarters. Or you may be virtually looked after by remote providers from headquarters, like the doc-in-the-box. You may even undergo surgery by an expert surgeon operating from headquarters over the Internet, controling robotic tools at your real location. Any or all of these possibilities are within reach; but, as we'll see in the next two chapters, we'll need concurrent advances in our ability to intelligently analyze and view medical data using smart software and visualization tools.

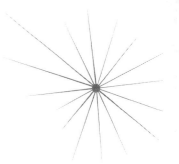

CHAPTER 3

ETERNAL VIGILANCE

There is nothing novel in the observation that we are barraged by an ever-growing variety of competitors for our attention in modern life, and information overload is a substantial problem in modern medicine. We accumulate more information than ever before. We see more patients in less time, and they're sicker. Doctors and nurses are constantly processing data about patients ranging from vital signs and lab studies to messages from patients and colleagues. Sometimes they are confronted with the complex demands presented by a single complicated patient, as during surgery or in the intensive care unit or emergency room. Sometimes the demands come from prioritizing issues in the management of a large number of patients in a practice or an inpatient service. Doctors are frequently presented with rapidly evolving problems wherein they must determine which information is most important, and what to do next, in a very short period of time.

In a 2000 *New Yorker* article Malcolm Gladwell distinguishes between different ways people react under acute pressure. John F. Kennedy Jr. panicked when he found himself in a darkening haze over Martha's Vineyard, and in his increasing turmoil he instinctively fell back on the instruments he understood best, his eyes and brain, whilst peering desperately out the window in search of a light. He failed to process the information on the dashboard instruments of the plane, which would have told him his was entering a death spiral dive. This same kind of problem can happen in the operating room.

For a variety of reasons, people often draw comparisons between anesthesiologists and pilots. One of the mottos of the anesthesiologist is "Eternal Vigilance," a quality equally important to a pilot. Instruments are a common theme, as is the concept that things can go from boredom to sheer hell in a very short instant. It's not surprising that many of my anesthesiologist friends are actually pilots as well. I had one of those sheer hell moments midway through my anesthesia residency, when things abruptly went from routine to panic. We were about to embark on the surgical hip replacement of an older woman named Ann Franzen. She had reluctantly decided to undergo surgery only after the hip had deteriorated to painful gristle. Orthopedists typically recommend that people live with the pain of hip arthritis as long as possible because replacement hips have a limited life span for reasons discussed in a later chapter.

In hip replacement surgery, the patient, once asleep, is turned on her side on the operating table, with the operative hip on top so that the orthopedists can get to the hip socket. Of the various ways to anesthetize a patient for this surgery, the commonly used method is general anesthesia; with the patient asleep, a breathing tube is inserted into the windpipe, or trachea. Ann Franzen was rolled into the operating room that morning and given general anesthesia for her surgery.

We placed her stretcher next to the operating table, and she moved painfully from the one to the other with a little help from everyone in the room. She lay on her back, and we talked about her kids as I put an intravenous catheter into a vessel in her arm, and applied a blood pressure cuff and electrocardiogram leads. Then I asked her to start breathing some pure oxygen through what people usually describe as a rubbery-smelling face mask. This is the point at which patients used to be asked to count backward from a hundred. Nowadays we usually just give an intravenous syringe full of a drug that puts people off to sleep, followed by a drug that paralyzes all of the muscles, including the critical ones used to breathe. The heart, of course, keeps beating, but all of the skeletal muscles are paralyzed, which helps the surgeons work and keeps the patient from coughing or gagging when we insert the breathing tube. When Ann Franzen signed up for general anesthesia, she essentially turned over the management of her critical life maintenance functions to me.

That day, I put the breathing tube into her trachea with no problem and took over responsibility for her breathing, which I could handle either manually by squeezing an anesthesia bag or automatically by turning on a respirator built into the anesthesia machine. After the tube was inserted, I squeezed the bag a couple of times, watched her chest rise and fall, and assumed that everything was in order. The next step was to roll her on her side so the surgery could begin. I started the automatic respirator so I could free up my hands, and the whole team of nurses, orthopedists and I grabbed different parts of her body—I had the head, a nurse handled the legs, someone rolled the chest, someone else the pelvis.

Like a lot of orthopedists, the doctor who handled the turning of the torso had done some weightlifting, and from doing dead lifts, he understood correctly that one squats and uses leg power to lift and avoid straining the back. The anesthesiologist typically directs "the turn" because he is handling the head and breathing tube, and so I indicated that we would roll on the count of three. The turn went smoothly, and the team immediately began to check various pressure points to make sure Franzen wasn't lying in such a way that a nerve would get pinched or that an appendage, such as an ear, was folded into an awkward position that could impede blood flow and damage it. Suddenly, several alarms started going off on the anesthesia machine. One indicated low pressure in the respirator, another low volume, and a third low oxygen in the patient. These alarms are comparable to the stall alarms in an airplane and are designed to warn the anesthesiologist just before the patient goes into a graveyard spiral.

Alarms go off with some frequency in the operating room, and the problem and its solution are usually readily apparent. I quickly concluded that all of the oxygen had leaked out of the breathing system somewhere, but then had to figure out where the leak was in very short order, because Franzen wasn't able to breathe on her own. The first step was to cut the automatic respirator out of the system and switch to manual breathing, squeeze the anesthesia bag and hope the problem was in the respirator, while sorting things out. However, the bag, which had been full of oxygen before we turned the patient, was now flaccid and empty and there was no gas in it with which to inflate the lungs. Increasingly frazzled because the easy answers weren't panning out, I quickly ran through all of the connections to see if anything was detached. Nothing was. I tried to use a back-up "flood" valve to inject some extra oxygen into the breathing circuit, but the bag remained flat, which meant that there was some catastrophic leak somewhere in the system that I couldn't find.

The problem was hugely complicated by the fact that the patient was on her side which, although helpful for the surgery, meant that I couldn't readily get at her mouth and airway. This is called the lateral decubitis position, and is one that most anesthesiologists don't like, for good reason. Like landing an airplane in the middle of a city, the anesthesiologist has a much smaller safety margin when caring for a patient in this position. At that point, I panicked.

I asked the others in the room to call for help, pulled the surgical mask down off of my mouth and started to blow air into her breathing tube to inflate her lungs. This was how my colleagues found me when they tumbled into the room seconds later. I was squatting down beside the table doing CPR through her breathing tube. While this awkward solution worked, in my panic I hadn't thought to use the backup oxygen and breathing system designed for just this eventuality.

With the help of the hoard of people who were now in the OR surrounding my blissfully unaware, anesthetized patient, we figured out what had happened. When the powerful orthopedic resident had squatted down to lift the patient, his left buttock came down on a lever that holds a canister of carbon-dioxide absorbing crystals in place in the breathing circuit, thereby dropping the canister and creating an enormous gap in the system. We lifted the lever back into place, things instantaneously returned to normal, and rest of the operation proceeded uneventfully. I never told Ann Franzen what happened, and I've never forgotten the feeling I had that moment.

In retrospect, the alarms did what they were supposed to do—they told me there was a problem—but they didn't give me enough information to do anything about it, because I didn't know where the problem was. It's like the night I got a phone call in the five-story Veteran's Hospital in Palo Alto, California. A nurse yelled into the phone, "The patient just had a cardiac arrest," and hung up, without telling me who or where. I knew there was a big problem somewhere in the building but didn't have enough information to do anything about it. Fortunately, smarter medical alarms and computer algorithms with the capacity to learn are on the horizon. The smart alarms of the future will sift through various data streams pertaining to individual, or groups of, patients and preprocess them for human doctors and nurses. They will get the right information to the right people, allowing them to provide the right care at the right time.

Just such an alarm had summoned my medical colleague Ben Kohl into a room where the bedside doctors were in the midst of doing a procedure on a patient. As mentioned, Ben was a doc-in-the-box that night and when he "peered" into the patient's room from the birds-eye vantage point of the ceiling camera, he realized the patient's chest wasn't moving. The alarm that had summoned him acted as a kind of rudimentary artificial intelligence, or smart alarm, virtually sitting inside the computer network looking for anomalies—in this case, the fact that the patient's blood oxygen level and heart rate were dropping.

Now admittedly, current smart alarms aren't *super* smart, they're just super reliable; and, unlike the nurse in the Palo Alto VA, they tell you something about the problem and who's having it. Smart alarms also never panic and they're eternally vigilant. If so programmed, they scan the instruments relentlessly: heart rate OK? blood pressure OK? breathing OK? heart rate OK? . . . Unlike humans, smart alarms don't get so caught up in a task that they lose track of the bigger picture. They're also smarter than the current alarms on individual medical instruments, which can only tell us that some threshold was just exceeded or something just became disconnected. Today's alarms are single-minded and know only about the instrument they're built into, and the best they can do is to tell a human something bad just happened. The smart alarms on tomorrow's medical machines will talk to one another, they'll learn, and they'll be able to tell a human when something bad is *about* to happen, where the problem is and what we can do about it. They're the perfect assistant to work with smart doctors and nurses who have a whole lot of things going on around them.

But what are smart alarms? How does a smart alarm differ from a regular alarm? At their root, all alarms are just algorithms; but whereas traditional alarms act like a reflex, with no thinking involved, smart alarms use smart algorithms designed to simulate human thinking, which is one type of artificial intelligence.

Artificial intelligence has been around in medicine for a long time, but the current proliferation of computerized patient information and the maturation of the field will force more widespread adoption of smart technology in medicine. The very first time most medical students are confronted with the product of a smart algorithm is the day they receive their assignments for residency.

The United States National Residency Matching Program, or NRMP, which has been around longer than the Internet itself, is an even more unusual matchmaking program than anything on today's Web. In their senior year, graduating American medical students create a rank-ordered list of residency programs indicating the places they'd like to train; the training programs, in turn, rank the medical student applicants they've interviewed. Both parties submit their lists to the NRMP in mid-January, at which point the approximately 30,000 students' preferences are dumped into the hopper of a computer along with the 25,000 available positions.

This NRMP matching computer can be thought of as the mechanical version of an old-fashioned matchmaker, and it has a kind of primitive artificial intelligence, perhaps at the level of a barnyard animal. After processing for three months, the results are made available both to the residents and the programs in sealed white envelopes on a fateful day in March. I was given a white envelope at 1:30 P.M. on Wednesday, March 16, 1983. At that moment, I was with all of my anxious medical school classmates in Dunlop Auditorium at the University of Pennsylvania Medical School, seated next to my then-girlfriend from the same class, on the aisle, about four-fifths of the way to the top.

The sealed envelope contained my name and that of the program where, after months of thought, the NRMP algorithm had decided I would spend the next three years of my life. Like my girlfriend, I had applied to programs on both coasts, and we had some imperfectly formed idea that we would end up on the same coast, perhaps close to one another. I generally preferred the Right Coast, and she the Left. We opened our envelopes at the same time, along with all of the rest of our classmates, and *voilà* . . . I was off to Stanford University on the leftward West Coast and she was headed to New York for our several-year-long indentureships. A bit over three months later, I arrived in Palo Alto to start my internal medicine residency in the first balmy days of July

1983. I rented a little ranch house on a postage-stamp-sized lot, with a back deck for sunbathing and, the big selling point, a genuine Californian wooden hot tub.

An internal medicine residency trains one how to take care of hospitalized patients, while, paradoxically, the typical practicing internist spends most of his career seeing patients in an office, and only limited time rounding in the hospital. Every Stanford internal medicine resident, however, had an office clinic through all three years of residency, where we learned to take care of outpatients while supervised by a faculty mentor. My faculty mentor was a thin man in his mid-thirties, with wispy blonde hair and a biting wit, named Ted Shortliffe. As I came to appreciate during my residency, Shortliffe was a very good internist. Much later I would learn that the clinic was only a sideline for him, and that he was primarily involved in something much more important.

Our academic medical clinic worked the way similar training clinics work the world over. At the start of the clinic, the receptionist would register a patient and the nurse would take him or her to a room, check the vital signs, find out what prompted the visit and record this information on a chart. When I worked my way to that patient's room, I'd grab the chart from a slot on the door, quickly peruse the data, knock and enter.

Over the course of my three years of training at Stanford, I went from being tentative and diffident, because I knew how much I didn't know, to being quick and efficient in my encounters. Because I was in training, however, I presented every patient I ever saw to Shortliffe. I'd finish my evaluation of the patient, walk into the clinic's alcove, sit down in a chair across from Shortliffe, who was usually leaning back in one of those squeaky office chairs with a stethoscope around his neck and a couple of pens squared up in the pocket of his white coat, and tell him what I'd found.

I knew he did something with computers when not at the clinic. Nevertheless, we didn't discuss computing much because there were lots of patients to see; we discussed the diagnoses and management issues of each one, which took a lot of time. Ted Shortliffe had a very logical approach to walking through each issue, and he could articulate his thought processes—his mental algorithms—very clearly and lucidly. I'd sit down, relate what the patient had said, what I found on exam, what I thought was going on and my plan. Shortliffe would occasionally catch some anomaly, we'd go into the patient's room together and he'd show me where I'd gone wrong with a hint of a grin.

I learned a lot from him about how to get from the occasional patient who complained "I have swollen nymph glands" to the diagnosis and treatment of what actually turned out to be mononucleosis. Over the three years I spent at Stanford, Ted Shortliffe and I saw each other for several hours weekly, but I never fully appreciated at the time we were working together that he was simultaneously developing into one of the pioneers in the field of what is now called medical informatics, and, more specifically, in the development of artificial intelligence in medicine—a field with roots extending back to the use of primitive computers by cryptanalysts at Bletchley Park in World War II.

In addition to being unaware of the ground-breaking work that Ted was doing on the mainframe computers adjacent to Stanford University Hospital, I was equally oblivious to the even more significant work that was going on in garages and buildings in and around Palo Alto. The San Francisco Bay area was evolving as a hotbed of personal computer development during this period: Xerox's Palo Alto Research Center, where the mouse and the graphical user interface were developed, was literally just around the corner, and Steve Jobs was working out of a garage right down the road.

As my residency at Stanford ended, I decided that internal medicine was not my calling and, although I finished my training in internal medicine, afterwards I returned to the University of Pennsylvania to pursue further training in anesthesiology and ultimately in critical care. Ted Shortliffe came to New York to be the chair of the Department of Biomedical Informatics at Columbia University, and we eventually met up again twenty years later when he gave a guest lecture for a medical informatics course I was teaching at the time.

Ted was educated at Harvard and Stanford, where he received his MD and Ph.D. His doctoral thesis, completed in the mid-1970s, was the development of a ground-breaking, computer-based program called MYCIN, which was designed to simulate the thinking processes of an expert infectious disease consultant. Computers themselves were relatively novel at the time, and the idea that an intelligent computer could take on some of the responsibilities of a doctor was both startling and intriguing to some, while threatening to others. Ted's premise was that general practitioners lacking a clear diagnosis might turn to MYCIN for help when confronted by a patient who looked and smelled as if he had an infection.

The term artificial intelligence has been much misused over the years and certainly should never really be applied, as I did in jest earlier, to the algorithms used by the NRMP. The most widely accepted IQ test for a computer is what's called the Turing test, named for Alan Turing, who is unequivocally one of the more interesting and tragic figures in modern science. Turing was a brilliant British mathematician, logician and cryptanalyst, and is widely considered to be the father of modern computing. He ultimately committed suicide after having lost his security clearance due to his homosexuality. He studied at the University of Cambridge and later at Princeton before World War II, during which he led one of the codebreaking teams at Bletchley that helped decipher German military communications. Turing's work laid the theoretical foundations for artificial intelligence; Ted Shortliffe was an intellectual descendant of Turing's and one of the early pioneers in applying his work to medicine—no discussion of machine intelligence would be complete without understanding the contributions of these two men. Based on the foundations of their earlier work, we are now seeing the deployment of intelligent computer algorithms, like the smart alarms I mentioned earlier, integrated into the daily practice of clinical medicine.

During his life, Turing led pioneering work in both the theory and mechanics of computer science, and proposed a still-relevant and objective intelligence test to see if a computer could act in a sufficiently lifelike manner to fool a human examiner. The Turing test of artificial intelligence is based on a human party game of the 1950s called the Imitation Game, in which a man and a woman are sent into separate closed rooms, and a third party acts as an interrogator who is only allowed to communicate with them through a series of questions passed under their doors. The interrogator's goal is to determine which room holds the male and which the female. The responses to his questions are typewritten to preclude identification by handwriting. According to the rules of the game, the woman's messages must always tell the truth, while the man's must always lie. Perhaps working through his own internal demons, Turing proposed a version of the game wherein a computer substituted for one of the closeted players. He suggested that a computer could only be termed intelligent when it could successfully deceive the interrogator for a period of time, thereby always lying and implicitly acting as the male according to the rules of the Imitation Game. Shortliffe conceived of a computer program that could play the role of a doctor.

Ted Shortliffe's Ph.D. thesis program, MYCIN, was an attempt to create a form of artificially intelligent medical specialist using a complicated but very rigorous set of rules, or algorithms, formatted something like this:

IF "patient age" < 7, THEN no tetracycline

As we've subsequently learned, there are a number of problematic features with the use of rules of this structure in medicine, which, in this instance, was intended to prevent the administration of tetracycline to children during the vulnerable period of tooth formation due to the permanent discoloration that tetracycline can cause in milk teeth. Doctors don't really use rules like this in their practice, because the medical world is too complicated for this kind of rule to work in every situation.

MYCIN was designed to be what is called an expert system, and it was specifically intended to act like an expert infectious disease specialist. In practice, the idea was that a nonexpert, perhaps an internist or pediatrician, would sit down at a computer terminal and answer a series of questions posed by the program about a problem patient. When the questions were all answered, the MYCIN program would provide several potential diagnoses ranked in order of probability, just like a human specialist. The program would indicate its degree of confidence in each of the probable diagnoses, show what rules led to each diagnosis (thereby explaining the program's reasoning) and recommend treatments. In other words, it acted a lot like I did when I presented a patient to Ted in the clinic. The fact that MYCIN didn't come up with just one right answer was good, because when most doctors think about a diagnostic problem, they tend to use these sorts of rank-ordered lists of possibilities, which we call differential diagnoses. Ted was also very smart to have MYCIN explain its reasoning, because this gave users a sense of how the program arrived at its conclusions.

MYCIN actually turned out to be a pretty good expert, all in all, generating correct diagnoses about 65 percent of the time when it was tested against real patients. While this percentage was less than the real human infectious disease experts given the same problems—the human experts were 80 percent accurate—it was a lot better than many doctors who were not infectious disease specialists. In spite of this success and similar proof-of-principle demonstrations in other areas of medicine, medical artificial intelligence has never threatened to replace real doctors, and certainly never passed Turing's test. It is only recently that new forms of artificial intelligence have begun to be integrated into real patient care.

In retrospect, one of the main problems that prevented the widespread adoption of MYCIN and other artificial intelligence programs of the era was their use of algorithms like the tetracycline rule. MYCIN had about 500 of these rules, all of which had the same general short-coming. If we take apart the tetracycline rule, for example, we can see that it makes a sharp boundary between the age that a patient can and can't get tetracycline; however we know that tetracycline doesn't just stop damaging teeth when a child blows out that seventh birthday candle. While the rules were intended to impose order, their structure was too crisp for a field in which a certain degree of elasticity is mandatory.

The rules in MYCIN were very good, but very inflexible, which created *another* problem. Because the rules were hard-coded, the computer would come up with *exactly* the same answer every time it was presented with the same problem. Mind you, there's nothing wrong with self-consistency in medicine, and you'd like to think a doctor's how-I-do-it algorithm doesn't vary with the amount of sleep she got the night before. But this kind of prescriptive rule-based program has no capacity to learn on its own; it therefore requires continuous rewriting whenever medicine changes, as with the development of new antibiotics or resistant bacteria.

Rule-based systems work great for residency matching programs, in which the following rules will always make sense:

IF Jack lists Stanford high on PREFERRED RESIDENCY
    PROGRAMS
AND, Stanford DOES NOT rank Jack on its list of DESIRED
    MEDICAL STUDENTS
THEN, Jack DOES NOT go to Stanford.

The same types of algorithms probably work pretty well in matchmaking Internet-based programs, where one can imagine that the programmers have codified rules based upon the personality profiles of prospective mates:

IF Jack ranks high on the BEER-PONG INDEX,
AND Jill ranks high on the CHINTZ INDEX,
THEN the LIKELIHOOD OF LONG TERM MARITAL
    SUCCESS will be < 80;
GO TO TRY AGAIN.

Very structured rules, however, are not necessarily well suited to a complex and rapidly changing field like medicine.

One thing that the tetracycline, residency matching and dating rules have in common is the fact that they are predeterministic; in other words, the programmers decided *ahead* of time what the rules were and wrote the computer code that way—none of the programs have a built-in feedback loop, as some humans do, that would allow them to get *better* based on experience. The lack of a feedback loop results in a system that is essentially analogous to a student who goes through school taking tests that never get graded and therefore at graduation has no idea whether she's really bright or really dumb.

Early expert computer systems of the 1980s, like the MYCIN program and a series of contemporaneous medically oriented computer programs, had no function by which to say "How did I do on that last problem and how can I do better on the next?" Nor did they have the capacity to deal with unexpected situations. At best, a mistyped entry might generate a message from the computer like "I don't understand that."

Since then, however, the field of artificial intelligence has evolved considerably, particularly in certain fields, as demonstrated by the consistent superiority of chess-playing computers over human grandmasters. To be fair, these chess computers rely to a large extent on what is called a brute-force strategy, in which they search mechanically through a huge set of alternative moves with extreme speed. Nevertheless, one of the developments underlying the improvements in artificial intelligence is the new field of *machine learning,* which encompasses a wide range of processes by which a computer improves its future performance based on its past experience—so that it doesn't keep making the same dumb mistake over and over again.

Turing predicted that it wouldn't be until the year 2000 that machines would become sufficiently smart, or perhaps deceitful, to fool thirty percent of a test's participants. Today's computers are very close, but in the opinion of experts no computer has truly passed yet. Of course, one could argue that our fact-filled medical school curricula doesn't necessarily teach our human students how to think and analyze intelligently either.

Returning briefly to the Turing definition of artificial intelligence, you will remember that in order to be successfully characterized as intelligent, the computer had to fool a human for more than five minutes. One of the things that quickly becomes apparent during medical school and residency is that there are certain book-smart students who excel at an activity we call roundsmanship. These students are able to regurgitate arcane medical references more or less on demand. For example, if a professor were to ask a group of medical students which animal is used to study leprosy, a roundsman would immediately exclaim "Armadillo!" Fortunately, while the rest of the students might end up feeling inferior each time a roundsman pipes up correctly, professors, having once been students themselves, are not necessarily fooled into thinking that the roundsman is *truly* intelligent. It turns out that it is actually not that difficult to *pretend* to be intelligent, as was demonstrated by another early computer project called ELIZA.

ELIZA was a computer program designed in the 1960s to emulate a psychotherapist. It was named after Eliza Doolittle, the Cockney girl who was taught how to appear more upper-class than she actually was. Its creator,

Joseph Weizenbaum, who was then a Massachusetts Institute of Technology faculty member, programmed ELIZA to behave like a classical Rogerian empathic psychotherapist, and intended ELIZA to be a parody of a human. However, Weizenbaum was much more successful than he anticipated, and many people who used the program believed that they were dealing with a real human being seated at the other end of the teletype link through which they communicated with ELIZA. A conversation might go like this:

> ELIZA: Hello, I am ELIZA. How can I help you?
> Me: I am feeling anxious.
> ELIZA: How long have you been feeling anxious?
> Me: For a month.
> ELIZA: Please go on.
> Me: I started feeling anxious when my editor told me I had to finish writing this book.
> ELIZA: Why do you say your editor told you that you had to finish writing this book?
> Me: Because that's what made me anxious.
> ELIZA: Do any other reasons come to mind?
> Me: Hmm . . .

As you might have guessed, this is a real conversation I recently engaged in with an on-line version of the ELIZA program, and now that you know ELIZA *is* a program, you can get a rough sense of how it might work. In essence, the program takes something I've said, tweaks it a little, and spins it back to me with an empathic flavor. You can imagine that those same folks who get sucked in by mail and computer scams might well believe that ELIZA was listening to them and cared deeply about their problem. When asked forty years after its creation why the program was able to fool many users, Weizenbaum said: "Let me compare this with a bartender . . . the drinking person projects his imagination about the bartender, who he is, what he knows. He might think, for example: 'This man has been a bartender for the past 20 years, he has heard a lot of stories and has developed deep insights' . . . but this is a projection."

The ELIZA experiment has spun off many imitators since 1966, some serious, some affectionate parodies; but not all of her offspring are academic endeavors. The Universal Life Church, physically headquartered in Modesto California, offers a simple on-line confession function for troubled souls—all you need to do is type in a compressed summary of your sins, check off the radio buttons indicating whether you've forgiven yourself and others, click the "Submit Confession" radio button and your slate is clean again.

Viewed from today's vantage point, MYCIN and ELIZA are very simple programs: MYCIN had about 500 rules, and ELIZA consisted of about 200 lines of programming code. Windows XP, by way of comparison, consists of 2 million lines of code. With that many lines it might even be possible today to develop a virtual marriage counselor. For better or worse, artificial intelligence has moved away from the creation of faithful, electronic reproductions of experts, or what might be called top-down programming, to a more bottom-up approach. Ted Shortliffe and Joseph Weizenbaum wrote every rule algorithm of their programs, and their creations were remarkable considering that period's computing power and data storage methods (prior to the development of stored program computers), but MYCIN and ELIZA were literally unable to think outside of the box. Just as the book-smart, medical student roundsman who knew about the armadillo might not turn out to be the best doctor, computer programs with a finite, prescribed set of possible behaviors will never replace an experienced clinician.

The key thing that distinguishes a really good medical student from a really good doctor can be summed up in one word—experience. The good doctor sees lots of people, with lots of diseases, and comes to learn the many ways a single disease can present in different patients as well as the many ways patients respond to illness. Their mental models of diseases and patients get better and better with experience, as do their algorithms for managing both.

Bottom-up artificial intelligence is possible today, where it wasn't thirty years ago, because of the massive amounts of data we now collect and store routinely. Modern computer systems can analyze information using various methods of machine learning and can "think" using an approach called inductive reasoning, in which observations and rules are derived from experience. In other words, they can learn. All animate life forms learn from experience. Even a flatworm will eventually scrunch itself up in advance of a light flash that it has learned to associate with an electric shock. After accumulating a database of two hundred or so simultaneous shocks and flashbulbs, the flatworm learns, unlike some Hollywood starlets, that flashbulbs are not always good. And the chemical program that the flatworm uses to burn this information into what passes for its brain is probably very similar to the programs we use in computing, since many types of machine learning algorithms were explicitly derived from natural processes. *Neural network algorithms,* for example, are bottom-up computer algorithms that mimic the behavior of brain cells. Similarly, *genetic algorithms* mimic natural selection. *Induction algorithms* develop rules based on experience. *Case-based reasoning* is a form of computer logic in which the computer draws on a database of previous cases and solutions to find an answer to a new case, just like an experienced doctor. *Fuzzy logic* is a mathematical system designed to deal with adjectives and adverbs, such as *feverish,* or *painful,* or *rapidly.*

While the term fuzzy logic may seem to be an oxymoron, it turns out that we all use machines that apply fuzzy rules to real-world tasks every day. For example, fuzzy-logic controls have been built into dishwashers, air conditioners, elevators, subway cars, automatic transmissions, antilock braking systems and image processing. Fuzzy logic allows a programmer to create a rule like:

IF "patient age" < YOUNG, THEN no tetracycline

While this rule may seem a little, well, fuzzy at first glance, it allows us to get around the problem with the MYCIN version of the same rule, in which the passage of a birthday flips the prescription rule overnight. In reality, the vast majority of the decisions we make and the actions we take in medicine involve fuzzy variables like "YOUNG." Take body temperature, for example. Every medical student learns the rule that a patient with a temperature greater than either 101.5 degrees Fahrenheit or 38.5 degrees Celsius has a fever; and when, and only when, a patient hits that temperature, the intern engages in a set of reflex actions including blood sampling, urine testing and chest x-rays . . . and sometimes even prescribes antibiotics.

But there are several problems with this FEVER rule. The first is that the metric and nonmetric versions of the rule aren't even exactly equivalent: 101.5

degrees Fahrenheit doesn't equal 38.5 C, it equals 38.6111111 degrees Centigrade, so the rule is internally inconsistent. A second problem is that human temperature varies considerably depending on where you take it: the mouth, rectum, skin, middle ear and bloodstream typically have different temperatures in the same patient at exactly the same moment. Something that seems like a crisp rule on the surface is actually inherently fuzzy.

Today's fuzzy antilock braking systems bring a vehicle to a stop far more quickly and with more steering control than the traditional foot-and-hand system by maintaining brake pressure between "Too Hard" and "Too Light" rather than allowing a human operator to slam on the brakes and send the vehicle into a skid. The rough medical equivalents to characterize degrees of anesthesia might be "Too Light" and "Too Deep."

When I take care of an anesthetized patient in the operating room, I am constantly monitoring heart rate, blood pressure, breathing and other variables, while simultaneously giving drugs in order to negotiate the middle ground between the Scylla and Charybdis of anesthesiology: "Awake under Anesthesia" and "Way Too Deep." It turns out that experimental fuzzy control devices similar to antilock brakes have been experimentally designed to administer anesthesia just about as well as I can. Fuzzy control of blood pressure, of mechanical ventilation and of a variety of other critical medical treatments are also feasible. And if fuzzy logic is good enough to control the brakes of an airliner with hundreds of people aboard, or a subway train full of commuters hurtling into a Japanese tube station, it's almost certainly good enough to automate the control of specific patient treatments in an operating room or intensive care unit in conjunction with human caregivers.

Neural networks, case-based reasoning and genetic algorithms are bottom-up, machine learning computer algorithms already in use in the nonmedical world to address a wide variety of problems. Neural networks, for example, can be used to detect motion, track people and track faces in security film footage and are being designed to identify terrorists in real-time airport camera footage. Similarly, the human expert at the other end of a computer help line may well be using case-based reasoning to find previous problems like yours, when the doohickey you just plugged into the laptop brought the system to a full stop. Genetic algorithms basically imitate natural selection to find near-optimal solutions to problems for which a large group of competing solutions exist. These Darwinian selection programs have been used by Wall Street mathematical wizards to develop extremely lucrative, but highly secretive, trading strategies; by industrial manufacturers to optimize scheduling; by Microsoft for optimal programming code performance; and by General Electric to design a better turbine.

Smart, self-teaching computer algorithms are equally well suited to medical applications and will be increasingly integrated into medicine as more medical data becomes computerized and more doctors and nurses chart and order treatments using computers. Intelligent algorithms can be used to remind the prescribing provider about allergies or potential adverse drug interactions. Case-based reasoning may help doctors in diagnosing rare or emerging diseases such as bird flu or bubonic plague. These same smart algorithms can be used to look for inconsistencies, patterns or errors in the way a student, intern or even an experienced physician practices medicine. For example, as I pointed out in my story about asthma and cat dander in the last chapter, asthma is often handled very differently in two different parts of the country or even by two practitioners in the same medical group. Smart genetic computer algorithms may help to identify the most efficient, cost-effective and safe ways to handle this common medical problem.

Automated medical image screening is directly analogous to the problem confronted by security experts looking for terrorists among millions of security camera images; and just as a system can be trained to recognize the facial characteristics of specific individuals using intelligent screening algorithms, it can be trained to identify pathology slides or x-rays that are abnormal and require further human scrutiny. So-called neurofuzzy systems combine the learning capabilities of neural networks with fuzzy rules and can automate control of medical devices. For example, a neurofuzzy insulin pump might determine when and by how much to increase and decrease an infusion in a patient with diabetes based on the stored data of fluctuations in that specific patient's glucose samples.

Back in 1984, when computer programs had to be short and there was very little in the way of computerized patient data, Ted Shortliffe and a colleague said, "Medical artificial intelligence programs are based on symbolic models of disease entities and their relationship to patient factors." Today, however, medical artificial intelligence programs and programmers no longer have to rely on symbolic models; they have access to enormous amounts of data acquired from real, rather than symbolic, patients with those same diseases. Modern programs can quickly learn about a disease's characteristics from a database of patients that a human physician might never see in a lifetime's worth of practice. Unlike experienced doctors, medical trainees, we say, tend to think of zebras when they hear hoof beats—they almost reflexively think of the most exotic explanation for a problem when they see a patient with a disease that lies outside their very limited previous experience.

I distinctly remember presenting a female patient to Shortliffe in our medical clinic. The woman looked jaundiced and had obvious liver disease; she couldn't walk straight and was a little addled. I interviewed and examined her and walked brightly into the conference room to tell him that I had stumbled upon a case of an extremely rare syndrome called Wilson's disease, in which copper accumulates in the liver and brain and has just those symptoms. As I went through the story, a small grin crept across his face, and when I concluded with my diagnosis, he said in an amused but still kind way: "She's probably just an alcoholic." And, of course, she was. He had the benefit of that universal leavening, time, which makes bread rise, good wine great and eager residents competent doctors. Like an experienced doctor, modern medical artificial intelligence algorithms will soon be able to distill experience from huge databases of patient encounters, to distinguish the hoof beats of a zebra from those of a horse.

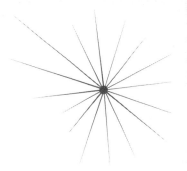

CHAPTER 4

A GOOD SKETCH

L ike most experts, doctors base their analysis and diagnoses of patients' problems on the mental maps they construct from many
sources of data, including vital signs, laboratory studies, physical
examinations and x-rays. As doctors get more experienced, their mental
maps become sharper and more detailed, and they arrive at an understanding of a problem more quickly and with greater accuracy. Their biggest
challenge is to identify critical trends in rapidly evolving situations; and
we're learning that pictures or graphics that integrate data into easily intelligible "maps" are a very good way to turn data into information.

Several years ago I was at an international meeting of anesthesiologists
in Barcelona. When the day's talks came to an end, we splintered off into
large table-sized groups and headed to restaurants for dinner. I ended up in a
restaurant overlooking the Mediterranean with a tableful of doctors from
various points of the compass; and as is often the case at gatherings of this
sort, whatever the business, we found common ground in talking about
work.

A British anesthesiologist told an unfortunate and unforgettable story that
evening about a case he had managed a few days earlier—probably because it
was still very fresh in his mind. Like most of us, he was an academic physician
and therefore supervised junior residents as they learned how to practice their
discipline. He and an anesthesia registrar (who would be called a resident in the
United States), had anesthetized a young man who had fallen off a scaffold and

fractured his thighbone, or femur. The patient had a bad break, with the bone protruding through the skin. He needed several hours of complicated surgery involving the placement of a long metal rod down the center of the bone. This is a very bloody operation, and unlike other surgeries in which most of the blood is suctioned into canisters so that everyone can keep track of the blood loss, orthopedic operations are often done with big incisions and lost blood can really be lost—into surgical drapes or the bedding or even onto the floor beneath the operating room table.

For operations on extremities such as the arm or knee, we usually use a tourniquet to prevent undue bleeding, but those on the hip or femur do not lend themselves to this technique. It is therefore critical in these operations that the anesthesiologist keep track of what's happening in the operative field as well as on the drapes and under the table. My colleague had left that OR to attend to other patients. A while later, he was urgently called back, and found the orthopedist performing chest compressions and the registrar frantically pumping blood into the IVs. The patient's heart had gone into fibrillation, and although the doctors worked on the patient for a full hour, they were never able to regain any blood pressure. They finally declared the young man dead.

After a case like this, there is invariably a grim period while the scrub nurses clean up. The patient's lifeless body lies on the OR table, and the doctors write code and death notes for the chart, explaining what they thought had happened and what they did in response. In this case, it wasn't completely clear to anyone why this young, otherwise healthy man with a leg fracture had died, and there was a lot of head scratching. My colleague was in the room during this period and, to his experienced eye, there was an awful lot of blood soaking the drapes and on the floor under the OR table. He talked the case over with the devastated registrar, who explained that he'd thought the patient was doing fine until the catastrophe.

My colleague had an idea, but wanted to work it through without seeming to second-guess the traumatized registrar, so he waited until everyone had left and the body was removed. He then returned to the room and activated a function in the anesthesia monitor that allowed him to see several vital sign trends on the same screen. He put the heart rate, blood pressure and blood oxygen trend graphs up simultaneously, showing the hour before the patient's death, and found just what he had expected.

In the half hour before the patient's sudden demise, the heart rate had accelerated—slowly and then more quickly. The blood pressure decreased over the same period and the difference between the upper value, or the systolic pressure, and the lower value, the diastolic pressure, had gotten smaller and smaller. So where the blood pressure thirty minutes before the patient's arrest was 140 over 70, and the heart rate was 70, the pressure had dropped to 70 over 50 and the rate had increased to 115 just before the patient's heart stopped. Finally, the blood oxygen was fine until minutes before the event, and then it started to head down quickly. None of this was nearly as obvious on the registrar's handwritten anesthesia record, which noted the trends less clearly; but shown together, the data trend lines showed the unmistakable record of a patient who was bleeding to death and whose heart had pumped faster and faster trying to compensate, eventually giving out. In most cases it's possible to turn this situation around even after a cardiac arrest, but once the heart goes into fibrillation, all bets are off.

Although the information was all there, the registrar was just too inexperienced or, perhaps, too inattentive to integrate it. My colleague quite correctly felt that he himself was partly responsible for not having supervised the registrar sufficiently. This was clearly an avoidable death, and human error played a big role. However, a large part of the problem in this case, and many like it, is the way in which medical data is traditionally handled in data-rich environments like the operating room and the intensive care unit. Taken in isolation, the blood pressure of 70 over 50 and the heart rate of 115 warranted concern, but didn't suggest the imminent catastrophe that was more readily apparent from the graphical representation of the trended data.

We have traditionally relied on humans to recognize trends in complex medical data; and we've used numbers or words to express trends that would be better displayed in pictures, where patterns can be recognized much more efficiently. This failure to discern critical patterns in complicated information is common to many fields, yet certain clear-thinking pioneers of every generation seem to be able to distill clarity from confusion. Charles Minard was a French civil engineer in the Napoleonic era. Like Stephen Hawking, who was able to clarify extremely complex concepts in his book *A Brief History of Time,* or Richard Feynman, nicknamed the "Great Explainer," Minard had a gift for distilling complex topics into simple, comprehensible pictures. He used those skills to create what some describe as the best map ever made: a two-dimensional depiction of Napoleon's ill-fated 1812–1813 march on Moscow. Minard's graphical map combines information about time, space, temperature, direction and manpower to show in a single, visually intuitive picture what happened to Napoleon's Grand Army of over 400,000 soldiers between June 24, 1812, the day it left Latvia heading east, and its ignominious return six months later at less than a tenth of its original size.

The map indicates the size and course of the army using a geographic track with a width proportionate to the size of the army. The daily temperatures during the march home are charted just below the map, allowing the viewer to see what the value was on a given day. As with the trend lines in the operating room, the story is told by a picture. The temperature was 0 degrees Centigrade on the day the army left Moscow and minus 30 degrees two months later. The army dwindled from 100,000 men to 8,000 during that westward march; the subzero temperatures during the November and December retreat took a much greater toll on the underdressed Grand Army survivors than any of the battles. The map tells the story of the campaign far more succinctly than any written record ever could. While a map lacks the flexibility of language, it can be an extremely efficient way to communicate complicated data. As it happens, Minard's map has a spare beauty as well.

Edward Tufte, a Yale University emeritus professor of statistics, uses Minard's graphic as an example of a masterpiece in his books and lectures on visual representation of complex information, and has called it "the best statistical graph ever drawn." Tufte's books can be found on the desks of innovative thinkers in almost every industry, because he has revolutionized approaches to efficient and effective communication in the age of computers. The problem of how best to display informative multidimensional data to humans, who typically think in three or four dimensions, is increasingly important as our ability to gather information grows exponentially. This is nowhere more true than in medicine.

The medical record has changed radically during the course of my twenty-year career. In the 1980s most information was handwritten, and the physical chart was the *only* record, by which I mean that there was one copy for each patient-doctor encounter. So, for example, there was a doctor's office record for each office visit and a separate hospital record for every hospitalization at every hospital a patient visited. Medical records were kept in file drawers or folders; if the record was sent to a doctor's office for review and got lost, or there was a fire, that record was gone forever. Doctor's notes had a narrative, sometimes breezy flavor. "Mr. Jones looks tip-top this morning. Great work by the nurses!" Or, as in an infamous note, penned fifteen minutes *after* a patient had died: "Mr. Smith looks lethargic today. Suggest he be mobilized to a chair."

Today's records are far more complex than they used to be for a variety of reasons. First of all, there is a lot more information. There are more tests to do. There are more regulations that generate more paperwork. There is more cover-your-ass documentation and testing for legal purposes. Today's medical records are just much more voluminous—which is not to say that they are better—in fact, in many ways, they're worse at conveying information. I am occasionally asked by a law firm or hospital to review medical cases in which something has gone wrong and they need an uninvolved expert to analyze the case. It's not unusual to receive several boxes of paper of hundreds of pages in the mail, and most of my work involves culling out the two or three critical pages of information holding the key facts.

Everyone recognizes that we need to computerize health information, and many elements of the record are already computerized, but electronic medical record technology has been slow to arrive. The interim solution we've taken at my hospital is to scan all of the paper notes into computers, anticipating the day that hospitals will become totally paperless. These scanned documents never get lost, and more than one person at more than one location can view the same record at the same time. Despite this, we still have a limited ability to search a record or a group of records for important information, because while the *picture* of the record is in the computer, the words are not in computer format. For example, we would have a very hard time searching scanned records to determine which of our patients received a particular drug (for example, Vioxx) that was only recently shown to have late, bad side effects. Scanned documents are also extremely inefficient forms of data representation. A single page of a scanned medical record typically holds no more than three or four meaningful pieces of information, many redundant observations and a lot of empty white space. As we'll see in a later chapter about nanotechnology, it is theoretically possible to put the entire contents of the *Encyclopedia Britannica* on a dot the size of a single punctuation point on one of these scanned pages.

Tufte uses the term "data density" as one way to characterize the efficiency of a display format. The Minard chart of Napoleon's campaign, for example, is data dense—it shows complex interactions and invites intuitive thinking. The chart encourages the observation that subfreezing temperatures had a profound impact on the French army during the November and December retreat, and the viewer can make this observation at a glance. Huge numbers of soldiers died from the cold, and it's very easy to imagine frozen cadavers strewn along hundreds and hundreds of miles of Russian roadsides. An average telephone book is another example of an extremely efficient, data-dense communication vehicle, with 18,000 characters per page. Compared to Minard's chart or the telephone book, a typical medical record is antediluvian. As an example of where Minard got it right a century and a half ago and we still have it wrong, let's compare his depiction of temperature as a trend plot with the way we depict the patient's pulse in the medical record.

More so than any other sign, the pulse has fascinated physicians since the dawn of medicine, and early Greek physicians such as Herophilus and Galen wrote extensively about pulses, classifying characteristics such as magnitude, speed, intensity and rhythm. The art of interpretation of the pulse was known as sphygmology, and through the centuries, a great deal of effort was expended on theories relating music and mathematics to variations in the pulse.

If one of the ancients were to leaf through one of today's medical records, they would find ample evidence that we still pay attention to the pulse—it is textually recorded diligently, almost slavishly, in almost every note throughout the modern record. Unfortunately, many of the wonderful classic descriptors such as "antlike," "gazellelike" or "wormlike" are gone. We record the pulse today as an unadorned, naked number. In fact, often a modern physician will just write "VSS," meaning "vital signs stable," which is about as informative as describing something as "OK." Galen wrote eighteen books on the topic of pulse interpretation and could be forgiven for scoffing at our ignorance—of course, today we have a deep understanding about the function of the heart and the arteries that wasn't available even a decade ago. We still care a lot about the pulse and the other vital signs—blood pressure, respiratory rate and temperature—but it's their behavior over time and their relationship to one another that matters.

Certain types of medical charts *do* show medical data in a graphical format so that it is possible to make inferences about the relationship of the data elements to one another. For example, the bedside flow-sheet maintained by nurses is sometimes recorded on graph paper and fever or pulse spikes can be seen at a glance. With careful attention to these graphs when they're available, you can often see events, like the increase in pulse rate and blood pressure that typically occurs when the surgeon cuts into the even well-anesthetized patient's skin with a sharp scalpel. The power of Minard-like graphical representations is their ability to convey information *visually*, in a way that humans are hard-wired to process. The observer can *see* the relationship of the ambient temperature to troop mortality or the blood pressure to the heart rate, rather than poring through columns of numbers and mentally recreating their trends.

Tufte proposes the use of what he calls sparklines, which are small, word-sized graphics or visualizations showing the temporal behavior of important data, such as vital signs. He points out that in the same space it takes to print "pulse: 105," it is possible to embed a graphic in the midst of the related text showing a visual trend line of the pulse, perhaps even with the normal range shaded. Sparklines are one example of a highly efficient way to distill large amounts of data into a very compact format; and the advent of high-resolution flat-panel displays has already changed medicine dramatically by enabling new ways of visualizing medical information. When I first started training, the only way to view an x-ray was to go to the radiology department and look at the photographic film stored there. If a film was lost, as it often was, we'd take another x-ray. If the film was underexposed, yet another. Radiologists actually had big magnifying glasses, like Sherlock Holmes is often depicted carrying, with which they looked at fine detail on the films. Today, however, the entire process is digital, and software lets the physician magnify, zoom, rotate and view data from any angle. Digital mammograms are an example of this new technology, and some studies have shown that the new technique is better at finding tumors than standard film-based mammograms. Moreover, digital images never get lost, and many people can view the same image from different locations at the same time.

The Nighthawk radiologists I described in the second chapter have such software on their computer workstations and use them routinely to enhance digital films. In fact, these tools are similar to ones spy agencies would have used not too long ago to analyze satellite photographs of enemy weapons, installations and movements. The tools these agencies use today are presumably even better. And while I had to use expensive high-resolution computer screens to view these computerized films not too long ago, the resolution of modern personal desktop monitors is now such that my desktop display is perfectly adequate to look at most films.

Chest x-rays are two-dimensional, but three-dimensional radiography is now routine as well, and a variety of imaging technologies can provide physicians with remarkably detailed reconstructions of virtually any body part. Scans are so detailed and so fast that we can distinguish anatomical features less than one half of one millimeter in size, and make highly detailed three-dimensional movies of the beating heart. These images are used to diagnose disease, steer surgery or guide the construction of tailored prostheses.

Stereotactic therapies use the coordinates of an anatomic location in space to precisely localize disease and, while not new, have become increasingly sophisticated as the resolution of three-dimensional imaging has improved with increasingly high-resolution CT and MRI scans. Stereotactic operations typically require two steps. The first step is the acquisition of a scan of the area of interest, be it a tumor or anatomic area, which is then used to develop a 3-D roadmap of the target area and its relationship to anatomic landmarks, like the tip of the nose or the ear canals. Step two is the actual procedure, in which interventions such as radiation, biopsy or microsurgery are applied to areas deep inside the body with great precision. Procedures that would once have been impossible today can be performed stereotactically using instruments passed along "safe" pathways through the body, and minimally invasive treatments for common neurological diseases are now possible as a result. Similarly, treatments like the proton beam radiation discussed in the first chapter, in which it is crucial to know the exact anatomic coordinates of a tumor, require detailed three-dimensional visualizations of the lesion in order to shape the proton beam and deliver the radiation with the right energy.

The treatment of Parkinson's disease, a crippling syndrome in which patients develop shaking, stiffness and slowed movements, has changed dramatically over the past several years. Drug treatment was the therapeutic mainstay as recently as ten years ago, but a variety of new, stereotactically based interventions have evolved as our understanding of the anatomic basis of the disease has improved.

The basal ganglia are anatomic switching stations in the center of the head; sensory, motor and traffic control is integrated and modulated in these critical parts of the brain. A variety of strange and unusual problems result from injuries to this relatively primitive portion of the brain. For example, diseases such as Tourette's syndrome, obsessive-compulsive disorder and stuttering appear to originate in the basal ganglia. The basal ganglia are also damaged in Lesch-Nyhan's syndrome in which afflicted patients engage in self-destructive behaviors such as finger-biting and head-banging. Foreign Accent Syndrome is a rare and unusual disease in which afflicted patients begin to speak their native language with the accent of a nonnative speaker. An American native, for example, might begin to speak with a British accent after having suffered an injury to the basal ganglia. The fact that these weird behaviors are directly associated with brain injuries rather than, as once thought, psychiatric illness, provides us with hints about the complexity and beauty of the human brain. Although this part of the brain was once a sort of neurosurgical Bermuda Triangle into which we feared to go, we can now intervene very precisely and knowledgeably, because we can visualize the anatomy.

Parkinson's can now be treated with three different, stereotactically based interventions that reach deep into the brain without the need for major brain surgery. These treatments are designed to turn off or destroy overactive portions of the brain that cause the disease's characteristic symptoms. With gamma knife treatment, radioactive beams are focused from several directions on a very small area in the basal ganglia, destroying the area with no need for an incision. Stereotactic *ablation* is performed by passing a small electrode through the brain into the critical portion of the basal ganglia; the tip of the probe is then heated and the target area cauterized. The most recent innovation is deep-brain *stimulation* using a pacemaker-like device attached to electrodes placed into the same critical part of the basal ganglia. The pacemaker can be programmed to block the overactive centers and, unlike the two alternatives, this treatment is nondestructive.

Stereotactic procedures are directly analogous to the use of satellite and image-guided military weapons, or to the GPS navigation systems that many of us use in our cars and boats. All of these technologies rely on a previously acquired map, or visualization, of an area of interest and then some subsequent live interaction with that map, be it precise electrode placement in some localized part of the brain, precision bomb targeting or GPS directions to within a few meters on a map. And, as with the Minard graphic, we are increasingly able to layer additional critical information from a variety of sources onto these maps to increase what Tufte describes as their data density and dimensionality. For example, several cell-phone-based applications permit me to create layered maps showing the traffic, ATMs, restaurants and cheap gas nearest my current location. Similarly, neurophysiologists can map activity in local parts of the brain onto their images to help guide electrode placement. The ultimate goal of information scientists like Tufte is the development of visually oriented displays that efficiently, comprehensively and accurately represent data in a format that humans can act upon—in other words, pictures that tell stories at a glance.

Heads-up displays, or HUDs, were developed for military and commercial aviation almost a quarter of a century ago to combine multiple sources of information into a single field of view. So that the pilot had no need to look away from the sky to see information about altitude, airspeed or other critical data, it projected onto the cockpit screen. HUDs have also been developed for auto racing, motorcycle helmets and skydiving, and the technology has matured to the point that we are beginning to see HUDs in high-end production cars. Several manufacturers have debuted technologies whereby information about speed, navigation and other instruments are projected onto automobile windshields making it unnecessary for the driver to look away from the road. Newer aviation HUDs have totally synthetic vision systems, integrating many dimensions of data, such as altitude, airspeed, heading, energy, longitude and latitude into a single integrated view of the current location of the aircraft and where it should be going. HUDs are now being designed as adjuncts to certain types of surgery, so that anatomic and physiology information from x-rays and monitors can be visually superimposed to give the surgeon a real-time view of the total environment.

While the merits of HUDs seem self-evident, a recent family holiday gathering gave me an unexpected glimpse of what medicine will look like in the near future. As is often the case, the adults were gathered in one area, and the children, including my three sons, had segregated themselves in a game room, where a four-player game of Halo, the science fiction video computer game, was underway. As you may know, Halo is the so-called killer application of computer gaming consoles. Parenthetically, my children consider me to be a troglodyte because of my adamant refusal to have a gaming system in my home and are ecstatic when they have the opportunity to play elsewhere. In fact, they seem to take particular joy in killing anything remotely resembling a troglodyte as it crosses their field of view in any of these games.

Each of the Halo players has his own quadrant of the television screen and operates a complicated, hand-held console with a lot of buttons. Their screen quadrant shows what amounts to an HUD of Halo's virtual world with life-force, power, proximity and ammunition indicators in the corners of the screen. In other words, they have a multi-layered, highly visual display that integrates multiple data streams about their avatars—their computer personas—and the environment into a single comprehensive picture. The speed with which the kids were able to play the game and synthesize the information was astounding and left me with the feeling that things were going way too quickly, as if I were someone who grew up driving a tractor on two-lane roads and suddenly found myself in the driver's seat of a car on a superhighway.

For those of us raised on Pong, the killer-application video game of the 1970s, watching kids compete in multiplayer Halo is very humbling. The only thing that made me feel any better was the fact that what I do in the operating room every day is actually very similar to playing Halo. As I sit at the head of a patient's bed during an operation, I am surrounded by anesthesia monitors showing lots of different, but related, pieces of critical patient data: EKG, blood flow, blood oxygen level, breathing rate and level of anesthesia. I have to monitor all of it while keeping a patient alive during a complicated operation. These are *real* life-force indicators. As I watched the children play Halo, it dawned on me that mental maps developed today in the gaming industry for kids will radically change the medical devices and display paradigms used by tomorrow's doctors.

Anesthesiologists, surgeons, cardiologists and radiologists—we call them proceduralists—are already using increasingly high-resolution displays that combine information about the patient on whom they're operating into at-a-glance pictures. Surgeons in training use surgical simulators and learn to operate on animated, virtual patients. The advent of new technologies such as the multitouch display surfaces developed by Microsoft and Apple will eventually revolutionize the way physicians interact with and visualize patient information. Surface computing allows a user to use natural hand gestures, using one or more fingers to point, drag, and draw or modify digital images. Doctors will be able to drag patient information onto computerized work tables, explore anatomical information, trend and graph vital signs and lab data, and navigate using visual metaphors.

In a typical office visit today, a clerk gathers up lab data and the patient's old records, and charts new vital signs into a record that the harried doctor snatches up as she enters your exam room. She leafs through the chart while you sit there wondering what lies within, runs through your current medical issues and does a physical exam, Then she dashes off a couple of written prescriptions. Finally, she scribbles or dictates a quick note for the chart.

In tomorrow's world, however, she will enter the room and drag your old history, graphical trends of your vital signs and lab studies as well as your most recent x-rays onto a computerized work surface for both of you to see. The work station won't have a mouse or a keyboard, although it may have a microphone for dictated material; she'll work primarily with her hands and fingers, opening and closing elements of your record with hand gestures and finger swipes across the surface of the display.

As you talk, you can both refer back to the display, on which, perhaps, she shows you the humps in the trends where your weight and blood pressure spiked higher a year ago, and how they've improved since you started your new exercise regimen. She may then chart your encounter as you watch, quickly and efficiently building the note with a series of hand movements. She drags the heart icon into the center of the screen with her hand, and then clicks it open with a single touch of the tip of her index finger. She'll swiftly touch her way through her findings, and then select the correct interventions. And she may dictate a brief addendum to the computer's microphone, which will automatically transcribe it into the record, to cover some unusual issue you've brought up. Finally, she'll send off the prescriptions with a final sweep of the finger to the pharmacy icon.

The scale of a visual display plays an interesting and critical role in the kind of information it can convey. Sometimes, it is important to have a comprehensive, wide-angle view to identify critical features of a landscape; whereas at other times, one may drill down to the details. For example, the bird's-eye view from a helicopter can tell a realtor certain information about a house and the surrounding property, but it still takes the eyes of an inspector who crawls around the foundations to determine whether the house has termites. Different kinds of information become visible as one zooms in and out, and this is nowhere more evident than in the current burst of discovery we're seeing with the analysis of the human genome. Genetic variants that increase the risk of specific diseases, ranging from prostate cancer to diabetes, from multiple sclerosis to heart attacks, from restless leg syndrome to gall bladder problems, have become apparent. A very powerful way to discover linkages between genetic patterns and disease is by displaying genetic information in such a way that a human can identify the patterns using the brain's hardwired pattern recognition capability.

I recently visited the Carl Icahn Laboratory at Princeton, where a lot of coordinated work is under way on genomics, the study of the relationships between genes and disease. The laboratory is a center for genetic research where computer scientists and what we call wet-laboratory researchers work side by side on decoding the genetic code that makes us all run. The faculty at this lab includes chemists, physicists, economists, computer scientists and biologists, all of whom focus on complicated problems having to do with biological phenomena.

Designed by Rafael Viñoly, the same architect who designed Penn's proton beam center, the Icahn lab building has a beautiful, airy, soaring atrium containing a sculpture of a whale large enough to hold several tables full of café diners, who eat in the belly of the beast as it were. The lobby also houses a large, matte-black, eight-by-fifteen-foot wall that could easily be mistaken for part of the infrastructure were it not for the fact that the walls around it are white. This black wall is actually a giant fifteen-megapixel computer display, used by the Princeton genomics researchers as a discovery space. The wall was designed by Kai Li, who is on the computer science faculty at Princeton; Olga Troyanskaya, another faculty member with appointments in both the computer science and genomics departments, has developed software for this massive, data-dense display space to discover genetic patterns that couldn't be seen on smaller-scale displays.

Troyanskaya and her graduate student Matt Hibbs demonstrated the display wall for me by showing a map comparing the activity of various genes from a huge yeast database. The map showed rows and columns colored in shades ranging from red to green; the red-labeled genes were relatively dormant and the green ones were overactive. The wall looked like a giant work of abstract Christmas art although there were clear, recurrent patterns. Hibbs exercised the software a little bit to show how the display could be used to mine information from the data and identify important sites on the genome. The utility of a wall-sized map of *yeast* genes might not seem obvious at first glance. But, imagine the same display showing genetic maps of *human* patients with prostate cancer and imagine that map revealing distinct bands of red and green indicating a series of overactive genes not seen with normal patients. Then imagine those observations leading to a treatment that could screen for or even cure prostate cancer. Genomic researchers all over the world are using advanced visualization and mapping techniques just like this to identify the footprints of genetically related diseases.

Information visualization is going to radically change the way hospitals, doctors and patients think about medical information, because as medical information has become increasingly computerized, it has also by necessity, become increasingly quantitative. Subjective statements like "Mr. Jones looks tip-top today," cannot be quantified and are, for better or worse, increasingly irrelevant in the modern medical record. But Mr. Jones' score of nine out of a possible ten on a hypothetical wellness scale is numeric and can, therefore, be tracked from day to day, visit to visit, or patient to patient. The use of a number rather than an adjective allows the patient and doctor to track progress over time. Similarly, when many patients' wellness scores are aggregated, it becomes possible for hospitals, employers, insurers and governmental agencies to track the wellness of patients across providers and therefore rate providers. Height, weight, pulse, blood pressure, hemoglobin, functionality and patient satisfaction are all things that can be quantified, and therefore tracked over time. The most efficient way for a human being, whether patient or provider, to evaluate the performance of those important medical variables is visually.

As more and more medical data becomes numeric, data visualization techniques become imperative because humans process visual information very efficiently. Whether you're a modern commuter driving down a super-highway at seventy miles an hour with hundreds of other drivers, too many of whom are pecking away on their cellphones, or a highly skilled aboriginal hunter in the rainforests of New Guinea, you are well equipped to process vast amounts of visual information. A skilled driver has learned to spot anomalous behaviors by other drivers extremely quickly but may not be so quick to spot food in the branches of the rainforest, while the reverse is true of the aborigine. Parenthetically, my own incompetence in a rainforest was very obvious on a recent trip to a national park in Costa Rica, where a native guide readily found bats, sloths and fowl that seemed completely invisible to me, even with the assistance of a telescope.

Intelligent preprocessing helps a good deal. Two excellent modern examples of powerful information displays adhering to the principles espoused by Minard and Tufte are *SmartMoney* magazine's Map of the Market and the Centers for Disease Control map showing the spread of obesity in the United States over the past twenty years. Both of these maps can be found on the Internet. The Map of the Market uses the colors green and red, just like the yeast genome map at Princeton, to display the financial market in a series of nested rectangles, with each rectangle representing a company. All of the little company rectangles are clustered in larger group rectangles showing what sector the company represents, such as financial, energy or health care. Large companies have proportionately bigger rectangles. The default map controls are set to display gainers in green and losers in red, and it is very easy to see, at a glance, how well the whole market, its sectors and individual companies have done over the past 24 hours or 52 weeks, depending on how you set the controls. Further, by clicking on a specific company's rectangle, you can quickly drill down to get much greater detail. On any given day, you can get a very quick overall view of how the market is doing, what sectors are performing well and which companies are outliers in either direction. You may then choose to see more information about a given organization. Imagine using this same technology to track the patients in an intensive care unit.

A simpler, but similarly informative animated data map shows the relentless, stepwise, year-to-year increase in obesity in the United States over the past twenty years, using colors ranging from blue to purple to indicate degrees of excess weight in each state. As one watches a visualization of the spread of obesity, it becomes very clear that the South and the Midwest tend to lead the rest of the country and that, while obesity was unusual in 1986, 20 percent or more of the population of virtually every state in the union is now substantially overweight. Again with this map, you can click on a state to zoom into much more detail.

Data visualization in clinical medicine is currently most widely used in fields such as radiology and client disciplines, such as neurosurgery and orthopedics, which use radiographic images to guide operative procedures. Virtual reality and simulation is a rapidly growing area used to train nurses and physicians in trauma management, cardiopulmonary resuscitation, team interaction and complex surgical procedures. I did a virtual laparoscopic gall bladder removal recently on a simulator with laparoscopic tools attached to a computer. Simulation technology is becoming so sophisticated that I could "feel" the virtual tissue I was "seeing" as I picked it up with laparoscopic forceps. Heads-up displays are built into robotic surgical tools, as we'll see in a later chapter, integrating data from medical monitors, radiographic data sets and the real-world view, to provide doctors with a wealth of layered information literally right in front of their eyes.

The proverb "A picture is worth a thousand words" is nowhere more true than in the modern world of medicine, where the sheer volume of data we acquire in daily patient care, research and public health defies description. Doctors show pictures to other doctors to explain diagnoses, to plan procedures and to teach. Doctors show pictures to patients as a way of bridging the gap in medical knowledge, and I've heard many patients express gratitude to their doctors for showing them a picture to explain their diagnosis. Pictures can also be a critical way to bridge knowledge gaps among many other disciplines.

The derivation of the proverb about a picture's value is uncertain, but it may have originated with Napoleon Bonaparte who said, "*Un bon croquis vaut mieux qu'un long discours*"—"A good sketch is better than a long speech." Minard's sketch succinctly described the time, place and causes of the death of hundreds of thousands of Napoleon's soldiers, much as simple trend lines of heart rate and blood pressure explained the death of the man described by my colleague in Barcelona. Fortunately, pictures have found positive uses in medical discovery, treatment and cure; our increasing ability to see, manipulate and interact with computer-based objects using touch screens will dramatically change many aspects of medicine.

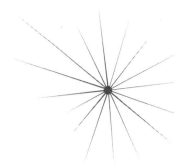

CHAPTER 5

SHADOW-PLAY

Surgery has changed dramatically since the days when it took two doctors to hold down the patient and a third one to wield the saw. A century ago the great American realist painter Thomas Eakins painted "The Agnew Clinic," which hangs in the hallway of my medical school. Agnew wears a white smock in his portrait, as do the doctors surrounding the patient. She is a young woman, and she lies on an operating table nude from the waist up. An anesthetist wafts chloroform over her face, while a surgical assistant is in the midst of a bloody left-sided mastectomy. The woman's pale, pink-tipped right breast is the painting's focal point, perfectly juxtaposing the innocent vulnerability of the breast with the implicit violence of the surgical act, set within the dispassionate objectivity of an educational venue. Agnew stands, holding a bloody scalpel in a pencil grip, in the foreground of a surgical amphitheater with tiers of students rising behind him. He exudes a sense of calm self-confidence as he makes a point to the rapt students behind him.

The surgical conditions depicted in the painting are unimaginable today: No one wears gloves or a mask and there is no hint of sterile drapes. This was, however, grand rounds 130 years ago, and it was quite aptly referred to as theater. Unfortunately, the day of surgical amphitheaters is long past, and today's grand rounds are a pale shadow of yore. Today, the expert may come with his PowerPoint slides on a flash drive, may hold a laser pointer to highlight text that zips on and off the screen from all different directions or fades in and out like the Cheshire cat. Having been on both sides of these modern presentations, I can assure you that the audience is half-asleep more than half of the time and never rapt.

Grand rounds at the Hospital of the University of Pennsylvania is often conducted in an amphitheater called Medical Alumni Hall, which has been around since the 1950s and isn't very grand at all. There is just a podium, dim fluorescent lights and a big sagging screen. The stage is raised, with a slightly curved proscenium. Rows of movie-theater-type, flip-down seats rise up from the floor of the hall in tiers; and, oddly, there is a brass plaque on many of the seat backs indicating that some donor has endowed it in perpetual memory of some former notable. The floor of the hall consists of a composite of white cement and black chunks shined to a high gloss, and each stair tread is taped with lines of antiskid tape. Fortunately, the hall does contain some acknowledgment of eras past. In contrast to the otherwise general blandness of the space, the left and right walls of the hall each hold a series of six portraits descending diagonally in ranks from the top of the hall to the stage level.

The portraits on the left are those of the most recent chairmen of medicine from the most current at the bottom to the most remote, five generations back, at the top; the portraits on the right wall are those of the chairmen of surgery, again in chronological order. If the room were higher and the rank longer, Agnew's portrait would have been high up in the back of the room because he was a former chair of surgery at Penn. One of the surgical portraits, midway down the steps, is of another giant in surgical medicine, Dr. Jonathan Rhoads, who died in 2002 in his nineties. Rhoads was one of the pioneers in understanding the importance of nutrition in surgical healing and, with others, developed intravenous feeding solutions.

Even in his late eighties, Rhoads still came to the hospital almost daily, a striking, stooped figure, still well over six feet in height, walking slowly on the way to his office. Like many surgeons, he carried his head jutting well forward of his body, a posture acquired from long years of operating with one's head sticking out over the patient on an operating table—and one that unfortunately predisposes many surgeons to degenerative disease in the neck. Rhoads loved the company of his fellow physicians and was married to one. He came to lunch in the doctor's dining area often, even well after retirement, where he might seem to be drowsing off over his soup until his eyes would twinkle out from under bushy eyebrows as he made some pithy, show-stopping comment.

The portraitist who painted Rhoads, while not in Eakins's league, was very capable and captured the patrician intelligence of the man, seated with his hands folded over his knees. Rhoads' hands are one of the most striking features of the portrait. They are enormous, beautiful, capable-looking hands—almost simian. As painted, his hands are as large as his head. The portraitist may have fashioned those big hands in the service of three-dimensional perspective or, more likely, as a masterful painter's statement about the man and what he did with those hands.

Because of their size, it is hard to imagine Rhoads' big hands actually doing anything meaningful inside of a human body, which is what one might naturally, but incorrectly, assume happens during surgery. While the surgeon does occasionally put a hand inside the belly to feel around for this and that, most surgery involves a trade-off between the size of the incision and surgical access to the part of the body being operated on, be it bowel, spleen, heart or brain. While a larger incision provides better access for the surgeon, there are a number of obvious downsides to big holes in the human body. From a purely cosmetic standpoint, big incisions leave unsightly scars. There is also an increased risk of infection, often greater internal scarring and more postoperative pain with large incisions. Small incisions, on the other hand, may force the surgeon to use longer instruments that can be clumsy to work with and limit visibility but enable him to operate externally. As a patient, you'd prefer to have as small an incision as possible; as a surgeon, you'd prefer to have a big open hole into which you could put your big hands as freely as possible. With the help of technology, the trade-offs are vanishing to the benefit of both parties.

It is instructive to compare the way the same operation would have been done when Rhoads operated with how it would be done today, and the Nissen antireflux fundoplication procedure is a perfect example. Named for its inventor, Dr. Rudolph Nissen, the operation is designed to prevent heartburn. A German-born surgeon who fled Germany before World War II, emigrated to Istanbul and came to the United States, where he became a faculty member at the Long Island College of Medicine, Nissen eventually finished his career in

Basel, Switzerland, where he died in 1981. Now you might think that heartburn, or acid reflux, shouldn't warrant surgery because pharmacy shelves are full of different medicines designed to treat it. However, this is a problem with which I have some personal experience and about which I can speak less theoretically than about many other medical issues. Mother Nature designed our digestive system very thoughtfully so as to extract as much nutritional value as possible from the foods we eat. The nose and tongue guide the brain to find food that smells and tastes good, and, at least until recently, is good for us. *Now*, of course, we've found ways to outwit the brain and have tarted up all sorts of toxins, carcinogens and nutritionally bereft substances to look and smell just like real food. Once the food, or its proxy, gets into our mouth, the teeth, the tongue and the salivary glands go to work to grind a mouthful into a paste, provided of course one chews long enough, which doesn't always happen in our fast-paced lives.

A chewed, or masticated, mouthful of food has already begun to be digested, because it's full of saliva that has already begun the process of breaking down starches and fat. Saliva is mostly water, but it also contains mucous, antibacterial compounds, bacteria and, most importantly, digestive enzymes; and, alarmingly to some, we humans make up to a liter a day of the stuff. Saliva also turns out to be alkaline (high) in pH, which helps to combat some of the bad effects of bacteria on teeth. Saliva flows into the mouth when food is present, and therefore the food paste that travels from the mouth down the esophagus is typically alkaline. The actual pH of a given bite obviously depends on the net interaction between the saliva and the food; certain foods, such as orange and tomato juices, are quite acidic to begin with and incompletely neutralized by the saliva, as those with heartburn are well aware.

The pharynx, which is the back of the throat, and the esophagus are passageways primarily designed to conduct the food you've just masticated into paste from the mouth toward the stomach. The esophagus is essentially a long, narrow, skin-covered muscular tube that milks the food into the stomach, even if you happen to be upside down when eating. At the bottom of the esophagus, between it and the stomach, there is typically a virtual valve, called the gastroesophageal sphincter, designed to isolate the esophagus from the stomach—and it is here that things get dodgy with heartburn.

Ordinarily, the alkaline food paste arrives at the gastroesophageal sphincter, which opens briefly allowing it to pass into the stomach where acid and a whole battery of more powerful enzymes await. These enzymes are designed to break the major food groups into molecules that will be absorbed by the intestines. The acid helps with this process and kills off any bacteria that may have come along for the ride. So the pH is high in the esophagus and then it plunges in the acidic environment of the stomach. The gastroesophageal sphincter is supposed to act like a sluice gate and keep the two environments separate, but the mechanism occasionally breaks down. It is pretty normal for pregnant women, for example, to have heartburn, because the pressure on the stomach is forced up by the growing uterus and child within. Later in pregnancy, the stomach gets so compressed that the sphincter is forced open, causing reflux of acid into the esophagus.

Other people have what is called a hiatal hernia, in which the stomach pushes partway into the chest, through the diaphragm, which in turn interferes with the function of the sphincter and causes reflux. A third group of heartburn sufferers have the problem for no obvious reason. In most cases in each of these groups, the problem tends to go away after a while; however, if it becomes chronic or severe, a variety of complications may occur. Chronic acid reflux into the esophagus can cause coughing, hoarseness or asthma due to acid traveling up the esophageal pipe and onto the vocal cords and into the lungs. Big meals make this more likely to occur, again because of increased pressure in the stomach, as does the recumbent position since acid is more likely to wash back into the esophagus when one is lying down. All of this could be written off as an admittedly very annoying inconvenience were it not for the fact that chronic reflux can also eventually result in cancer of the esophagus—probably because of the chronic irritation and inflammation from the acid.

I had the second kind of reflux—the one caused by a hiatal hernia—and did everything I could to avoid surgery. I tried antacids like Tums. I tried histamine blockers like Tagamet and Zantac. I tried proton pump inhibitors like Prilosec and Prevacid and Nexium. I had the head of my bed propped up six inches higher than the foot to prevent the reflux of recumbency, which prompted a friend and his wife, who stayed in that bed one night as our house guests, to point out that I should have told them to bring pitons—they both looked a little drawn the following morning. None of these interventions worked, and I developed a chronic throaty whisper and irritating cough. Finally, I saw a gastroenterologist who suggested a reflux test and had a look at my esophagus with a scope. Both tests suggested that it was time for surgery, and the antireflux procedure of choice is the Nissen fundoplication.

Dr. Nissen's operation involves wrapping the top half of the stomach around the lower end of the esophagus to create, in effect, a new gastroesophageal valve that prevents the acidic stomach contents from flowing backward. The surgical juncture between the stomach and the esophagus is tucked up under the breastbone, and it is difficult to get at, even through a large open incision. In Dr. Rhoads' era, the 1960s through the 1980s, the surgeon would have made a long incision in the top left portion of the belly wall, along and just under the rib-cage, an area where a lot of muscle attaches.

In order to access the stomach and esophagus to do Nissen's procedure, to hold the bowels out of the way, the surgeon would use retractors that come in two versions. The first, at an academic medical center like mine, is human: a junior assisting member of the surgical team, often a medical student or intern, is tasked with holding the bowel out of the way. The senior surgeon positions the blade of a hand-held metal retractor, which is long and curved, and used like a hoe to hold abdominal contents out of the way. He then transfers the metal handle to the assistant, typically with the admonition "Don't move"—I use the word "handle" generously here, because the grip of hand-held abdominal retractors is often merely a flat extension of the blade, lacking rounded, finger-friendly engineering. It was often necessary and very painful to hold the retractor with some force for several hours during an open Nissen procedure. Of course, like the little Spartan boy, human medical-student retractors never complain.

The second form of retractor is mechanical. Developed by a thoracic surgeon, this device is called the Bookwalter self-retaining retractor, and it looks like a big, misshapen cog-wheel. The wheel is first clamped to a post that sticks up from the surgical table so that the cogwheel forms a circle over and around the operative area. Self-retaining blades are then attached to the cogs on the wheel after having been positioned to hold the abdominal contents out of the way. This minimizes the need for additional human assistance, and a skilled surgeon can operate pretty efficiently without a lot of help this way. Needless to say, Dr. Bookwalter is much admired among junior surgical interns.

Once the retractors were optimally positioned for what is called surgical exposure during the open Nissen, the surgeon would bend his head out over the operating table and go to work. When he was done, the wound was sewn up, and the patient would undergo several weeks of recuperation and agony with the slightest movement of the stomach muscles, which had been sewn back onto the rib-cage. This was my first option, but a brand new approach was under development at the time I needed surgery.

Today's newer version of the Nissen fundoplication looks very different from the outside: It is performed laparoscopically using scopes and instruments passed through small holes in the abdominal wall. An abdominal laparoscopic procedure involves the following steps. First, a small incision is made in the patient's belly, and carbon dioxide is used to inflate the inside of the abdomen, creating a space between the abdominal wall and the gut, which consists of the stomach, bowel and other intra-abdominal organs. The operating room table is then tilted so that the patient's head is up and the feet are down, and the bowel settles down into the lower, pelvic portion of the abdomen like wet pasta in a bowl. This leaves an air bubble around the top of the stomach and the lower part of the esophagus, which is where the action takes place in a Nissen fundoplication.

Once the abdomen is inflated with gas, the surgeon passes a sterile videoscope through another small hole in the belly wall to display the inside of the abdomen on a television screen. This allows him to see the surgical field without bending over it, albeit without the depth perception provided by two eyes. Using a variety of laparoscopic instruments, the surgeon operates while monitoring his work on a television screen. The surgeon's end of the instrument has scissorlike finger grips, some of which are designed to cut and cauterize at the same time, while others are used to hold tissue or sutures.

When the laparoscopic procedure is done, the surgeon needs merely to suck the gas from the abdomen and sew up the small surgical portholes. Once I had decided to go ahead and have the Nissen operation, I sat down with my surgeon friend who was going to do the procedure. We talked through both options, the open and closed versions, and I decided to go with the laparoscopic version which, while admittedly still new, had the advantages of less time in the hospital afterward, a presumably shorter and less uncomfortable recovery and a lower likelihood of infection. The surgery went smoothly, although I was much more debilitated afterwards than I expected given the fact that the only surface evidence of the operation was three small half-inch scars. I came away from the experience with a much clearer sense of the concept that you can't judge a surgical "book" by its cover. The big operation was under the skin, and I needed a couple of weeks to feel fully recovered. However, the fatigue did not prevent me from returning to work much more quickly than I would have had I had the traditional version of the surgery. I am really glad the option was available. Now, I rarely, if ever, have heartburn; and when I do, it goes away with minimal treatment. Newer techniques may eventually eliminate the need for any surgery at all.

I have been an anesthesiologist during the transitional years when surgeons converted from open operations to laparoscopic versions of the same procedure, and watching some surgeons get used to laparoscopic instruments was akin to watching someone eat with chopsticks for the first time. Interestingly, some doctors who are very deft when operating with their own hands and eyes are still unable to function as effectively with laparoscopic tools.

The benefits of laparoscopic surgery to the patient are obvious: small incisions, less post-operative pain, less scarring both internally and externally; and a significantly decreased risk of infection. For these reasons, to the extent that we are safely able to perform surgery using less-invasive techniques, surgical techniques are inexorably moving in that direction. Similarly, minimally invasive catheter-based procedures such as angioplasty have replaced big operations, such as many instances of coronary artery bypass grafting. Even newer techniques such as aortic stents inserted through catheters in the groin have replaced large surgical procedures in which the patient's whole abdomen is opened to access the aorta from the outside in order to sew in a graft. Laparoscopic surgical techniques have been designed for plastic surgery (such as breast augmentation), abdominal surgery and heart and brain surgery. The next generation of advances will use robotics.

The Department of Surgery at my hospital recently invited heart surgeon and robotic surgery pioneer Dr. Randall Chitwood to showcase cutting-edge procedures in this successful new field. Chitwood is internationally renowned, and surgeons came from all over the world to Greenville, North Carolina, to learn from him how to operate with a robot. On a film he shows of the procedure you can see two metallic pincer hands operating in synchrony to pass and receive a suture through heart tissue. Each robotic appendage has just two digits, a thumb and a finger. Even though they are controlled by a skilled human surgeon, there is a distinctly alien quality to the crab-craw-like way they move.

The human hand has more than twenty degrees of freedom, and it can be cast into any of an endless variety of shapes in shadow-play on a back-lit screen. It is an instrument with remarkable dexterity. To give you a sense of what a degree of freedom means, an object that can only move back and forth along one axis has just one degree of freedom, whereas rotation around that same axis represents a second degree of freedom. A drill bit chewing its way into a piece of wood is operating with two degrees of freedom, forward-back, and clockwise-counterclockwise. When a surgeon passes a curved suture through tissue, she is rotating her wrist in one of its degrees of freedom; and the movements of the wrist are analogous to those of an aircraft, with pitch, yaw and roll. The wrist's pitch is its up and down flapping movement. The yaw is the side-to-side movement one might make when silently telling someone to "Stop!" The roll is like the twist of the wrist made by a surgeon rotating a surgical instrument to "drive" a needle through tissue.

Laparoscopic surgery represented a huge step forward, but it actually came at the expense of surgical degrees of freedom. Laparoscopic instruments essentially constrain the multidimensional freedom of the human hand into a lower dimensional space. For example, a laparoscopic needle driver can only get to a specific site in the abdomen along one axis, the one from the hole it came through to that site. In the corresponding open procedure through a large incision, the surgeon can move her arm, wrist and hand around much more freely to pass a needle more readily from a variety of angles. Robotic surgery, however, has all of the advantages of minimally invasive laparoscopic surgery without constraining a surgeon's degrees of freedom.

When Dr. Chitwood operates robotically on the mitral valve of the heart, he makes a tiny incision between the ribs on the right side of the chest, then passes a camera, light source and two robotic hands through the ribs into the chest. The anesthesiologist has already collapsed the right lung to allow the robotic instruments access to the right side of the heart. The robotic hands have lots of joints, and therefore many degrees of freedom; they're like little insect appendages. Unlike human hands, robotic appendages can operate around corners, as I saw first-hand on a recent visit to an operating room in my own hospital.

I walked into the operating room to watch a completely new application of robotics in head and neck surgery. The scene was surreal. The room was laid out in an unusual way in that the patient was lying with his head toward the OR door and the anesthesiologist was at his feet—usually the patient is the other way around. This unusual set-up allows the surgeon free access to the patient's head which is critical for cancer operations like this one, which was being done to remove an almost certainly smoking-related cancer of the vocal cords.

The patient was on his back on the table with a metal tongue retractor holding his mouth open—imagine someone sitting behind you, asking you to tilt your head back, inserting a tongue depressor and asking you to say "Ahhhhh." Four instruments darted rapidly in and out of the patient's mouth. Two were suckers managed by a female surgical resident, and the other two moved seemingly of their own accord on mechanical arms. The automaton hands angled in from both sides around the resident, like crab appendages, and flitted back and forth, with sudden abrupt changes in angles, making buzzing mechanical whirs. There were intermittent sucking sounds; and an occasional puff of smoke emanated from the patient's open mouth. Four or five flat-screen televisions arrayed around the room showed what was actually going on in the back of the throat, where I could see sucker tips and grippers working in a coordinated dance to grab, cut, burn and suck things behind the patient's tongue. The mechanical arms seemed to be preoccupied in their work, but, positioned as she was between them, the resident seemed to be at some risk should their intentions change.

I stood there mesmerized for several seconds before I heard someone bark out from a corner of the room: "Suck that blood." Looking around, I found the man behind the machine, the surgical Wizard of Oz, Dr. Gregory Weinstein, sitting in the corner of the operating room at a large, humped grey console, where he peered through a pair of goggles and rapidly manipulated both hands and feet, like a church organist playing a complicated fugue. His hands controlled the surgical mechanical arms moving several feet away in the patient's mouth, and his feet controlled a clutch, the cautery and the camera with separate pedals. He was in the midst of an operation called a partial laryngectomy, the surgical removal of a cancerous portion of the vocal cords, that would previously have required much longer, much more disfiguring surgery. He invited me to sit down at the console and look through the goggles.

I sat down briefly in his seat to peer into the camera's eye view of the operating field, and the robotic software automatically adjusted the eyepiece width and focus to give me a three-dimensional view of the larynx. The robotic graspers looked very much like creatures from movie versions of *The War of the Worlds;* I could see where they had nipped away most of one of the vocal cords.

Compare the standard operation to remove part of the larynx with the way Weinstein does it robotically. The nonrobotic surgery requires an incision that essentially cuts the face in half to get at the cancer. The robotic operation is done through the mouth. All else being equal, which would you choose?

Dr. Weinstein doesn't even have to wash his hands to do this operation, because the robotic hands are actually doing the work and they're very clean. He just saunters in, sits down in a comfortable chair at the robotic console, scoots the chair forward and starts to operate. He sees the operative area in three dimensions through a set of eyepieces. His fingers slip into scissorlike finger controls. The robotic instruments can be quickly swapped from cutters to graspers to suture drivers. There is even a voice controller with which the surgeon can direct the position of the camera.

Each of the interchangeable robotic surgical hands is digitally controlled by its own computer program. The image the surgeon sees is digitally filtered to eliminate background artifact. The temperature of the camera's lens is digitally controlled to prevent fogging. These features make it possible to operate safely inside the body's cavities through little holes, and as with laparoscopic surgery, carbon dioxide can be pumped into the body to make space where there isn't normally a cavity. Surgeries are becoming less and less invasive, and there's no need for the robot console or even the surgeon to be in the same room as the patient. Provided the right network connections are in place, a surgeon can operate on a patient across or under an ocean, and perhaps even in outer space.

In addition to the ability to operate through small holes, the digital nature of robotic surgery provides other capabilities. For example, the robot can be programmed to eliminate tremor. While surgeons are supposed to have steady hands, many don't, and the robot can filter the tremor out of a surgeon's movement, which can be extremely useful in delicate applications where fine control is critical, such as sewing small vessels or nerves together. The robot can also be programmed to scale movements, typically down, so that a one-centimeter movement of the surgeon's fingers is translated, for example, to a quarter-centimeter movement of the robotic hand. This is obviously very useful in microsurgical procedures in the eye or on microscopic nerves and vessels.

One major shortcoming of today's robotic devices is the lack of sensory feedback: The surgeon can't really feel the tissue he's operating on and must rely on experience and visual cues to know how much he can stretch a given tissue before it tears. However, the solution to this, called haptics, is on the horizon. Haptics is the science of engineering tactile sensation into computerized applications. For example, haptic joysticks and computer mice use vibration or changes in resistance to movement to give the operator an artificial sense of texture or gravity. A haptically enabled surgical robot will give the surgeon a sense of the elasticity and weight of the tissue she's picking up, cutting or driving a suture through.

There are other drawbacks to today's surgical robots. There is a steep learning curve for surgeons, the operations often take longer that their nonrobotic equivalents, the robots are large and heavy, and the machines are very expensive. On the flip side, postoperative stays for patients are shorter than those for standard operations and the associated cost-savings can justify the initial investment. Parenthetically, many new surgical devices start out big and expensive and rapidly become smaller and cheaper. For example, when I was in training, we were one of the few centers with what was called a lithotripter, which was developed to break up kidney stones. The lithotripter of yesterday was a tank of water the size of a small swimming pool into which we lowered fully anesthetized patients on a gantry, which is a very dicey procedure; today's devices use gel pads that are pasted onto the patient's back, and they're much smaller and cheaper.

Today's robots are large and unwieldy, but the robots of tomorrow will radically change the way we think about surgery. A British group has just announced plans to develop what they call the i-Snake, a multi-articulated robotic tube with many degrees of freedom. With an approximate diameter of a penny, the device is designed to snake its way through body tubes—such as blood vessels, airways and bowels—to some critical location and then deploy instruments that will operate from within, leaving no external scars at all. If successful, this technology will almost certainly be smaller and simpler to use than today's mastodon surgical robots. There are already preliminary animal studies of operations done on organs inside the abdomen in which the entire operation is performed through an incision on the inside of the stomach. The modern laparoscopic Nissen I underwent a decade ago will soon be antiquated by one of these newer robotic approaches.

In the coming decade, telemedical, robotic surgeries will become far more common. Dr. Weinstein was seated at a console next to the patient, but there is no technical reason that he couldn't perform that same surgery from his office, from another city, or, for that matter, from a ski lodge in Colorado. To be sure, remote robotic surgery requires skilled assistants at the patient's side for preparation and backup, as well as what's called a guarantee of service (or GoS) from Internet network providers to make sure the network doesn't freeze up mid-stitch. But robotic telemedicine provides a way to extend the geographic reach of highly skilled specialists to locations that might not need those services on a routine basis. Today, highly skilled specialists make missionary trips to medically underserved countries and hold yearly clinics where they operate on children with facial deformities or congenital heart defects; tomorrow, these children may not need to wait for those services. The specialist will operate remotely and as needed in conjunction with local surgeons, and thereby provide more immediate care.

Surgery is only one of the fields in which robots are entering medicine. Surgical robots represent high-end, *tour-de-force* technology; there are a lot of other jobs around a hospital for which robots are ideally suited and could thereby free the hands of people whose skills would be more effectively employed elsewhere.

For example, at the Hospital of the University of Pennsylvania, we have deployed a cute little R2-D2 knock-off named Tug, who has no hands, unlike his facile, surgical big brother. Tug has taken over a job formerly performed by a pharmacy technician, who already has plenty to do; Tug travels tirelessly from the basement pharmacy supply area to deliver drugs to patient floors all over the hospital.

You might think that the lack of hands would be a significant handicap for a messenger-bot, and that Tug would be better off as a mendicant rather than a medic; but he's treated like a little prince in our hospital where humans can wait for an eternity to get an elevator. I recently shadowed Tug as he carried drug supplies from the basement to a satellite pharmacy two buildings and six floors away, and here's what happened on the trip. Tug was loaded up with drugs by human tenders in the main pharmacy, although there were no narcotics for obvious reasons (the poor little guy has no way to defend himself). Once he was fully loaded, he set off on his own and rolled down the corridor to the elevator, where he sent out a radio call to the effect that he needed a ride. His summons was transmitted wirelessly to the hospital network, which identified an unemployed elevator—one in which no buttons had been pushed by humans for a period of time—and sent it to the basement. The doors opened and he smugly rolled into it alone. Tug *has* to be alone on the elevator because he executes a complicated movement to turn around within it during the trip—unlike R2-D2 his head doesn't spin and his eyes only have one degree of freedom.

When the elevator arrived at the floor Tug requested, he soundlessly exited and rolled down the hall with his drugs. He doesn't say much unless someone gets in his way, and even then he's polite, intoning "Waiting to proceed," in a bland, synthesized Midwestern American voice, almost exactly like the one used by Stephen Hawking's synthesizer. When Tug arrived at the intensive care unit to deliver his drugs, I assumed he'd be stuck because, lacking arms, he can't push the door control button—but he's smarter than that. He radioed the network again, which opened them for him—one machine's courtesy to another. Tug makes rounds like this all day and all night long.

I recently attended a demonstration of another mobile medical robot that can be used to allow a doctor or nurse to engage in virtual patient rounds. While this may sound inhuman to some, this approach is one way for providers to extend their reach when geographic or temporal barriers would otherwise prevent a visit. I know of patients in nursing homes who are only able to see a doctor when he makes occasional rounds at the facility or when things have become sufficiently dire that a trip to the hospital is necessary. Like Tug and the surgical robot, the rounding robot comes equipped with a camera for navigation. The robot has a collision-avoidance system and, using it, a physician can virtually roll into a patient's room and engage in a two-way conversation in which the physician can see the patient on his TV screen while the patient can see her on the robot's screen. The device even has a stethoscope, although I didn't ask if it was kept at 98.6 degrees Fahrenheit.

Robots will eventually be developed for other unskilled medical jobs. Like home robotic devices developed to clean and wash floors, pools and gutters, hospital versions will be designed to clean and, if necessary, sterilize corridors, operating rooms and instruments. Similarly, robots have been designed to hand instruments to surgeons, and robots can help more skilled providers in some aspects of patient care.

Hospitalized patients too often complain about the amount of time it takes for a nurse to answer a call for assistance; the nurses complain that they are overextended in caring for ever-sicker patients. Patients may need medications or assistance getting in and out of bed. These are tasks where robots can assist in the future, allowing the nurses to attend to the tasks that humans do best. The Japanese have actually developed an admittedly primitive robotic nursing assistant designed to help lift a patient. It can be seen in a remarkable video clip on the Web. The robot is a large, wheeled, minty-green machine with spatulate paws that responds to commands spoken by a human nurse who says: "Please hold up that woman," pointing at a patient wearing a pink bathrobe on a bed.

The robot, which is apparently able to see as well as hear, looks around carefully and points one of his spatulas at the patient, saying: "The woman on the bed?"

The nurse answers "Yes."

Lacking imagination, a weapon, or a perhaps simply due to the limited worldview of newly created beings, the robot ignores the ambiguity in the command "hold up" and responds "I understand."

It proceeds to roll over, very gently scoop both spatulas under the woman and pick her up. It then returns to its original position, intoning mechanically, "I have succeeded in holding."

Indeed he had, although, were I there, I wouldn't have had the heart to tell him that most of the rooms in my hospital have recently been equipped with overhead hydraulic hoists so that our human nurses no longer have to man-handle our increasingly obese patients.

The hands of past surgeons were remarkable instruments, and far more el-egant than those of the nursing robot's spatulas, the surgical robot's graspers and, of course, poor Tug's stumps. With our hands and minds, we've designed tools that mimic, translate and extend our own hands and minds. Robotic tools and techniques will proliferate in the future of medicine in ways that we're only beginning to imagine. Within a decade, Tug will no longer be the only lit-tle prince in the hallway, nor will he be the object of curiosity. Robotic drones will clean the corridors ceaselessly. Robots will transport mail, medical equip-ment and blood products. And robots will almost certainly be used to a much greater degree in the operating room, as devices get smaller, less expensive and easier to use. Above all, robots will free the warm hands of humans to better care for other humans in need.

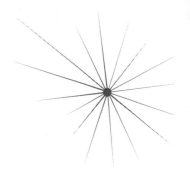

CHAPTER 6

THE THINKABLE

On January 8, 1942, in the midst of World War II, a baby boy was born to a biologist and his wife in England. The first few years of his life were spent in London and punctuated by German bombs, sirens and blackouts. His parents adhered to the prevalent philosophy of the time, which allowed them to persevere in the face of rations and capricious death: They believed in luck. As his mother puts it, "*Luck . . .* We have been very *lucky . . .* we've had some disasters, but the *point* is we have survived. *Everybody* has disasters, and yet some people disappear and are never seen again." She has high cheekbones, and her facial muscles have been sculpted by decades of working against, and defeating, gravity and disasters. When she talks about disaster, she sits a little straighter, tilts up her chin and draws in a breath through her nose before proceeding.

She raised her son as normally as possible during the war, pushing him around in a pram when it was safe to be out and about. The child had what his aunt described as a very large head and pink face; and he was apparently always in motion, both with his body and his animated face. He was a bright child, a bit of a monkey and liked games. He told his sister he had found eleven ways into their house in North London, several of which involved climbs up gutters, over portions of the roof and into windows.

While his parents were on a trip to India, he stayed with his aunt and uncle and took a sudden fancy to Scottish dancing, an athletic form of solo dance that evolved in the Highlands and is usually performed to bagpipe music.

Pictures from his youth show him astride a horse, standing next to a bicycle; later, in his college years, he's clearly the clown, grinning impishly and leaping in the air with arms spread wide. He waves a handkerchief in one photo and it's his face that leaps from the crowd of more serious undergraduates in every shot. He was also the coxswain for the varsity crew team, a job requiring a sense of rhythm and the ability to whip eight oarsmen along an exhausting kilometers-long course.

At one point in his late teens, he went to a park with his mother and younger brother, and while there, climbed a tree, which necessitated some tricky handwork on one branch, although certainly not anything more challenging than some of his expeditions into his house. His mother later concluded "He was testing himself out, I think . . . I didn't realize."

She said, "I think he began to notice (at that point) that his hands were less useful than they had been."

Not too long after, he fell down a flight of stone steps, bouncing all the way to the bottom. He lost his memory for a bit, and his first question on coming to was "Who am I." After a few hours he was able to reconstruct the days, then the hours and, finally, the minutes prior to his fall.

In the winter of his twenty-first year, it was quite cold, and many of the ponds iced up. He and his family went skating in a "not very advanced way"; he fell. And he couldn't get up. With the help of his family, he eventually managed to get off the ice and his mother took him to a cafe to warm up. Decades later she recounts a disastrous moment, the kind she so correctly indicates we will all go through. Her voice breaks and her usually quiet fingers fidget on the arm of her spindle-back chair. She says "he told me then . . . all about it"— what he described was his progressive loss of control over his body.

"It was diagnosed," she said.

After a series of tests, including a myelogram—a frequently painful test in which dye is injected into the spinal fluid and the patient is tilted this way and that by a radiologist—he was told he had motor neuron disease.

His mother, as one might imagine, had questions, and the neurological consultant agreed to entertain her visit "in a rather grand way," although, as she put it, he was evidently surprised she'd "bother to come round . . . I mean after all I'm only (his) mother."

The consultant allowed as how "It's all very sad . . . brilliant young man cut off in the prime of things . . ."

His mother pressed him: "Well *what* can we *do?* Can we get physiotherapy? Can we get *anything* like that . . . that will help in any way?"

The consultant responded, "Well . . . actually no. There's nothing you can do. More or less . . . that's it."

At the age of 21, he was given two and a half years to live and told "straight and flat" that he was gradually going to lose the use of his body, and eventually, "only his heart and his lungs would be operating . . . and his brain." He was also told that "eventually he would have the body of a cabbage, but his mind would still be in perfect working order" and that at that point "he'd be unable to communicate with the rest of the world."

The young man went into a profound depression, from which he was only rescued by a young woman he met and married shortly after—an example of what his mother describes as his "luck." A picture of the wedding party shows him propped off-kilter on a cane next to his bride. His hand is tightly entwined in hers as if the cane and her hand are the only two struts preventing gravity from having its way with him. Unlike the photographs of his college years, he looks deadly serious, as do his bride, her parents and his father. His mother is the only member of the party not looking at the cameraman. She is looking directly at her son with an enigmatic smile.

Eventually he came to "flatly accept that this (death) is what was going to happen to him." And only then, after previously coasting through school and life, "he started to do some work." A friend described a visit for dinner at the young couple's house, at the conclusion of which the man headed off to bed, hauling himself laboriously up each step to the bedroom on the second floor, using every baluster along the way, chipping the paint off the risers with his less and less useful feet. This nightly ritual was his way to forestall the inevitable wasting of his muscles.

As the doctors had predicted, by the age of 23 his strength and health had deteriorated to the point that he contracted pneumonia while on a visit to the European Organization for Nuclear Research and consequently was admitted to the intensive care unit of a hospital. He recovered, but it became necessary during that hospital stay to insert a tracheostomy into his windpipe to help him breathe, which meant that he could no longer talk, again as foretold.

Despite the predictions, however, he did find a way to communicate, although one that was extraordinarily cumbersome. The hospital had a clear plastic panel for patients on a respirator, upon which were painted letters of the alphabet arranged according to their frequency in the English language. With a little practice, he, his friends and family learned the use of the device. They would sit opposite one another and hold the panel up between their face and his, and then track his eyes as he looked from letter to letter to spell out one word at a time. This painfully slow method of dictation had become necessary because, just as he'd lost the ability to talk, his handwriting had deteriorated to the point that it was essentially illegible.

The last words he ever wrote were those of his signature. He signed an entry in a book entitled *Admission to Office*, a registry of the signatures of every university teaching officer at the school where, by dint of his brilliance and because of the work that he'd started to do only after being given a 30-month lease on life 16 years earlier, he had become chairman of Mathematics. The registry was kept as a record of the lineage of academic chairmen at Cambridge. The signature reads "S. W. Hawking"; the previous line reads "admission to the office of Lucasian Chairman of Mathematics." He signed the book in 1979, almost 30 years ago. Just as the rank of portraits of medical departmental chairmen in my hospital shows the lineage between Agnew and Rhoads, the Cambridge book shows Hawking's signature as well as that of each of his predecessors extending back to Isaac Newton, who held the Lucasian Chair between 1669 and 1702.

Hawking wrote *A Brief History of Time* in 1988, a best-selling book about the Big Bang, black holes and string theory, and the subject of a subsequent movie released in 1991 with the same title. Hawking still works on the physics of black holes, which are collapsed stars so dense that their gravitational force prevents the escape of light. As his neurologist predicted 45 years ago now, he is so disabled by amyotrophic lateral sclerosis, or Lou Gehrig's disease, that he is unable to talk and can communicate only using a voice synthesizer. Still, the work he has done in those years has significantly changed our understanding of the universe.

Lou Gehrig's disease affects the nerve cells that control voluntary muscles and leaves the involuntary muscles of the bladder and digestive tract untouched, sparing the senses of sight, hearing, smell, taste and touch. However, the muscles that control the mouth and vocal cords *are* affected, often quite early, and speech becomes slurred or unintelligible—hence the need for an artificial voice.

In a laborious process, Hawking now uses a finger-activated plate to control the construction of sentences on his wheelchair-mounted laptop computer. The sentence text is then parsed by a voice synthesizer and converted to sound. He manages about 15 words a minute, quite a bit less than the 150 words one hears in a typical minute of a book-on-tape. Of course, 15 of Hawking's words are probably worth a lot more than all of the prattle most of us utter in a lifetime.

An even more extreme example of a creative mind constrained by limits on communication was depicted in the 2007 film *The Diving Bell and the Butterfly*, based on the memoir with the same title by French journalist Jean-Dominique Bauby. Bauby suffered a massive stroke that left him in a state known as the locked-in syndrome, in which the only muscles an affected individual can control are those of the eyelids. Bauby wrote the entire book, which describes his daily life, by blinking his left eyelid when an assistant running through a frequency-ordered alphabet hit the correct next letter. It took an average of about two minutes per word to write. Ten days after the highly successful publication of his book, Bauby died.

The original voice synthesizer Hawking used was designed in the 1980s by the Digital Electronics Corporation using technology invented at the Massachusetts Institute of Technology. The device was of American manufacture and could speak with several male voices, including those of Huge Harry, Perfect Paul, or Frail Frank, all of which have an American accent. When first forced by his progressive disability to use a synthesizer, Hawking chose the voice of Perfect Paul, understandably preferring it to the two alternatives. Today, many of us would quickly recognize the voice as Hawking's, provided, of course, the subject was not the weather. The U.S. National Weather Service's automated radio-broadcast forecasts were announced using Perfect Paul's voice through the late 1990s. Sadly, Paul was finally replaced by another synthesizer named Craig in 2002, who won a run-off with Art, Linda and Donna, three alternative synthesized voices. Perhaps panicking under pressure, Paul had evidently became hard to comprehend when broadcasting storm warnings; Craig supposedly keeps a cooler head.

Voice synthesis has progressed dramatically since the development of Hawking's original device in the 1980s, and it wouldn't be surprising to find that there is a plummy Cambridge Don's voice in a software library somewhere. I'd also bet that Hawking, who has maintained his wit along with his wits, would shy away from it. Of his current voice, Hawking says: "it varies the intonation, and doesn't speak like a Dalek. The only trouble is that it gives me an American accent." For those of you who don't watch the BBC television series *Dr. Who,* Daleks are a fictional extraterrestrial race of mutants.

The Lucasian Chair of Mathematics at Cambridge was founded in 1663, and the list of notable previous Lucasian professors includes the Nobel Prize–winner Paul Dirac, one of the founders of quantum mechanics, and Sir Isaac Newton. Hawking, in some sense, owes his voice to the man who held the Lucasian chair between 1828 and 1839—Charles Babbage. A curmudgeonly mathematical genius, Babbage was the intellectual founder of the programmable computer. Babbage's Difference Engine was designed to replace the human computers who calculated mathematical tables essential to the conduct of what was, at that point, an enormous British Empire. The merits of Babbage's mechanical design were fully evident to his contemporaries, and he was, as a result, well funded to create a working version. Unfortunately, the professor was more of a thinker than an executor. While he did manage to start construction on what would eventually be a 15-ton, steam-powered calculator, he never got it to work. The first functional model of his machine had to wait until 1990, when it was shown by modern computer researchers at the London Science Museum to be both feasible with the engineering tolerances of the 1800s and accurate in calculating remainders from division to more than thirty decimal points. In effect, Babbage's 15-ton machine would have been able to what our hand-held calculators can do today.

The principles Babbage used for the design of the Difference Engine were later used to plan his Analytical Engine, which, while also never built during his lifetime, would have been programmed with punched cards like the ones described in an earlier chapter that were used in the 1970s. Today's digital computer that synthesizes Perfect Paul's voice for Stephen Hawking is a direct mechanical descendant of Babbage's Analytical Engine, just as Hawking is Babbage's direct descendant in the lineage of Lucasian chairs.

Today, Hawking is able to use the Internet, make phone calls, open and close doors in his home and operate lights, music and television, all from his wheelchair. Moreover, the newly constructed Cambridge Centre for Mathematical Sciences was designed with provisions to allow him to move about the building virtually unassisted. In the movie about his life, Hawking is seen sitting in a suit in his automated wheelchair, clicking sentences into his synthesizer. The chair has a red-cushioned support as a headrest. The touch-activated plate with which he sometimes interacts with the computer sits in his lap. There aren't many books in his office, but there are posters of Marilyn Monroe. Hawking says, at one point, that because of his inability to write down equations, he decided to work in an area of physics in which geometric shapes were meaningful.

Aside from the physical constraints imposed on him by Lou Gehrig's disease, Hawking's major handicap could be characterized as a limitation in the bandwidth with which he is able to communicate his ideas. More specifically, the speed with which Hawking can communicate his thoughts is limited by the constraints of the text parser and voice synthesizer. He has to build every sentence word by word and phrase by phrase before it is sent to the synthesizer for vocalization. He can, however, absorb information with his eyes and ears just as well as the rest of us. His communication pathways are therefore asymmetric like many home network connections, which are designed to download Internet Web pages much more quickly than they upload. Both Hawking and the average home Internet user have great inbound bandwidth, but asymmetrical limitations on outbound communication.

When watching Hawking in an interview, one becomes very aware of the slow pace of his speech and the long pauses between a question and his response while he builds the answer with his computer. While he may not speak like a Dalek, it's as if he were speaking from a slightly slower time-space continuum—or perhaps as if the words were escaping from a black hole. Yet, however odd this appears to be, this is exactly analogous to the way all of us communicate with the Internet every day, from the computer's perspective. Whether you're using a mouse, a keyboard or even a net-cam, the much-faster-thinking computer is sitting there idly while you, the seemingly slow-thinking human, construct what it is you want to do or say.

The bandwidth with which you download your thoughts into the computer is substantially lower, and therefore slower, than the bandwidth you use when communicating with other human beings in a real-world setting because the human-computer interface is much less efficient than the human-to-human interface. The good news is that, while the latter may even be getting less nuanced due to the advent of computerized video-gaming devices and text messaging, the former are improving by leaps and bounds. Hawking, for example, has now replaced the finger plate he once used with a much faster eye-driven mouse. The new device tracks his eye motions and moves the mouse accordingly.

Watching Hawking at work on a computer has always reminded me of my first programming adventures. I tried to explain how they worked to my children at dinner one night. I concluded that the best way to do so would be with a good sketch. They love computer games, so I described Zork, the first game I ever played on a computer, created, like Hawking's synthesizer, by MIT students. The game was designed for mainframe computers, and it used what is known as a command line interface—all text, no graphics. The game started with the following opening gambit from the computer:

> You are standing in an open field west of a white house with a
>   boarded front door.
> There is a mailbox here.
>   >

The ">" from the computer was the prompt for me to respond. The game didn't come with directions. It just started . . . like that. Eventually, after a series of missteps, I learned that if I typed "E" the computer understood "go East," "D" meant "go Down," and "O" could open many exciting doors behind which there were thieves, ogres, swords, mazes and lots of treasure. "H" turned out to be the motherlode, because it was the Help function. I got months of entertainment from that game. I told my children about it at dinner, and at least one of the three of them actually made a scoffing kind of noise, as if to say "How could something so dumb have ever been fun." They were, after all, skilled at playing Halo, the graphically robust video computer game. I took that scoffing noise to be a thrown gauntlet and within short order was able to find an online version of Zork, which they all ended up loving and still play.

The only difference between the Zork I played back in 1977 and the version I showed my kids was that I had used a teletype interfaced through what was called a modem to a stand-alone mainframe computer. My kids use one of the several laptops in our house, all of which are connected over a high-speed interface through our high-speed home network, to the Internet, and somewhere out there to the computer containing the modern version of Zork. The speed of the connection of my home laptop to the Internet is 1.741 megabits per second; 30 years ago I played the game at a mere 300 bits per second—four orders of magnitude slower.

In order to connect to the remote computer back in the 1970s, I picked up an old-fashioned phone and dialed the number to the mainframe computer's modem pool. When it was answered, I plugged the mouth and earpieces of my handset into what was called an acoustic coupler, which had two rubber sockets and was designed to screen out the background noise in the room. The coupler allowed my terminal to whisper to the mainframe; but the connection was so slow that I could watch the resulting letters creep across the screen one by one:

"Y . . . o . . . u . . . a . . . r . . . e . . . s . . . t . . . a . . . n . . . d . . . i . . . n . . . g . . . "

In fact, sometimes, the link was so slow that it only managed 15 words a minute, like Hawking, who would have eked out this sentence from the first paragraph of *A Brief History of Time:*

"Y . . . o . . . u . . . 'r . . . e . . . . v . . . e . . . r . . . y . . . . c . . . l . . . e . . . v . . . e . . . r . . . . y . . . o . . . u . . . n . . . g . . . . m . . . a . . . n, . . . v . . . e . . . r . . . y . . . c . . . l . . . e . . . v . . . e . . . r" said the old lady. "B . . . u . . . t . . . . i . . . t' . . . s . . . . t . . . u . . . r . . . t . . . l . . . e . . . s . . . . a . . . l . . . l . . . . t . . . h . . . e . . . . w . . . a . . . y . . . . d . . . o . . . w . . . n!"

Internet connections are now so *fast* that entire screens pop up in less than the blink of an eye. Network connections have become this fast because we've developed special wiring, such as optical fibers, and hardware, such as network adapters, to increase the bandwidth of communications. Keep in mind that most of us are currently much more interested in *inbound* speed, which is the speed with which our computer screen is displayed, because our typical *outbound* transaction with the Internet involves just a single, uploaded mouse click requesting a whole screen's worth of downloaded data—a Web page's worth—from a remote server somewhere on the Internet. In effect, every time we click on a Web page hyperlink, we effectively win a jackpot's worth of computer bits that flood onto our screen, painting the new page to which we've just browsed.

Returning to Stephen Hawking for a moment, think of that dazzling brilliance impatiently pent up by the 15 words per minute chokepoint between his brain and his manuscript. How nice would it be if that quirky, funny sentence about the old lady and the turtles from *A Brief History of Time* were able to flow more trippingly out of Hawking's text parser or off his tongue?

Reassuringly, it turns out that a lot of people are working hard on finding ways to make that happen. A variety of researchers are working on *direct* interfaces from the brain to electronic devices such as wheelchairs, synthesizers and computers. Although the primary focus of these researchers is to find ways to help the handicapped, a not-so-little side benefit of any solution that provides a direct interface from the brain to a computer is that it can be used to enhance the speed of communications between a normal person and any other computer-enabled device, perhaps a car or a spaceship.

A Massachusetts company founded by a prominent brain–computer interface researcher from Brown University, John Donoghue, has developed technology for direct mind control of computers using a microchip implanted on a portion of the brain. Donoghue won the 2007 Zulch prize for his work, Germany's highest honor for neurological research and one previously awarded to Stanley Prusiner, the Nobel Prize–winning discoverer of the agent that causes many of the neurological wasting diseases such as the so-called mad cow disease. Donoghue's Braingate chip projects 100 gold wires over the motor cortex of the brain, the area that controls our movements. The chip is thereby able to directly intercept neural signals originating in the motor cortex that are intended to control walking, typing or the movement of a mouse. The chip is designed to allow patients with severe motor disabilities, like Hawking's, to directly control electronic devices such as computers and wheelchairs with their thoughts.

Another mind-control researcher, Michael Callahan, is a recent graduate of the systems and entrepreneurial engineering program at the University of Illinois. He has developed an externally worn device called the Audeo that translates signals from the brain directly into speech. Unlike the Braingate chip that requires surgical implantation, the Audeo can be worn around the neck of a patient and is designed to intercept the neural signals from the brain on their way to the vocal cords and the muscles that control the formation of sound. Callahan and his colleagues have translated these captured speech thoughts so that they can either be synthesized directly to speech or used to control a wheelchair with commands like "E," "F" and "R" for East, Forward and Reverse.

Callahan formed a new company to develop the wheelchair-control and speech synthesis technology and has partnered with National Instruments Company, which sells laboratory software. I recently saw a film clip of the Audeo at work, in which Callahan and his partner, Thomas Coleman, demonstrate the capabilities of the technology.

The clip starts with Callahan standing next to Coleman, who is in a wheelchair. They're in a dimly lit hallway of what looks like an old school building. Callahan is tall and thin, and his shirttails hang out. He looks very young, and the video looks like a high school science project. Coleman sits next to Callahan with steepled fingers, his left leg crossed over his right, which is propped up on one of the wheelchair's foot supports. Callahan introduces what turns out to be a remarkable clip, saying "Alright. We're going to demonstrate the Mind Control Wheelchair." For the next two minutes, a mute, motionless Coleman cruises up and down the hallway on the chair, executing complicated turns, starts and stops, all without moving a finger.

Coleman later describes how he does it, saying "People often ask whether I just *think* where I want to go . . . when really you have to have some kind of *intent* behind it. It's a step *above* pure thought, but a step below actually moving or speaking." At a recent National Instruments meeting,

Callahan demonstrated the ability of a random, untrained attendee to "think" a sentence onto a screen using the same technology—the sentence was "This is really neat stuff."

The term subvocal speech has been used to describe the technology that Callahan has developed, and it has a wide array of potential applications. NASA is currently working with subvocal speech recognition for astronauts as they work on the International Space Station or while on spacewalks. The same approach could be used for secure communications over cellphones or during military and security operations. And you may eventually use it to talk to your computer. Both the Braingate neural implant and the Audeo have direct applicability to a variety of disabilities such as Lou Gehrig's disease, strokes and spinal cord injuries. These technologies can bypass damaged nerves and muscles, allowing direct control of devices that can move them from place to place, control the user's environment or communicate with computers. The implications for the rest of us are enormous.

As the safety and accuracy of these neuroelectronic devices grows, so too will their bandwidth. To put this in perspective, imagine how long it took for Hawking to write *A Brief History of Time* without benefit of a keyboard and how much easier things might have been if he had had a direct mental interface to the computer. Of course, it may well be that in this case brevity was the mother of attention. Hawking was evidently told by an editor that his readership would halve with every formula he included in the book, so he only used one—Einstein's famous one relating energy, mass and the speed of light.

The personal computer on which I am currently writing spends millions of idle central processing unit cycles virtually tapping its fingers while waiting for me to click each word into the word processor. The minute I am able to *think* my thoughts into a word processor, the keyboard will become a relic. I can say this with some certainty because, unlike many of my supple-fingered friends, I painfully hunt and peck.

It's a pretty safe bet that one way or the other, whether because of the impending revolution in touch-enabled devices or interfaces like the Audeo, the keyboard will go the way of the rotary phone within the next decade. I wasn't really that surprised to find out recently that one of my cellphone-equipped 12-year-old sons had no idea what a rotary phone was. This same son tends, sometimes for better, sometimes for worse, to take the shortest possible path between thought and speech; and, for better or worse, technology is allowing us to shorten that pathway to a greater degree with each passing day.

Our ability to extract meaningful information from nerve signals, such as the ones intercepted by the Audeo on the way to the larynx, or to translate sounds, sights and touch into the kind of information nerves understand, is growing exponentially. There are a variety of other neuroelectronic interface devices under development or already in regular clinical use. Prosthetic neuro-electronic ears are now routinely implanted in operating rooms every day to re-store hearing comprehension for patients with certain kinds of deafness. Bionic vision is in its infancy, but retinal implants have been developed and de-ployed in patients blinded by diseases like retinitis pigmentosa. Admittedly the field of vision of these devices is only a few pixels (the separate units that con-struct an image on a television or computer screen), but a few pixels of sight is enough to guide the blind along a line on the pavement. And, all of these tech-nologies are in the early stages of development.

Bionic prosthetic limbs are the focus of a substantial, government-funded effort due in large part to the number of soldiers undergoing amputations fol-lowing injuries in Iraq and Afghanistan. One of the predictable but unfortunate byproducts of every war are advances in medicine, and a variety of enabling technologies are maturing in parallel and presage explosive advances in neuro-electronics over the next few years. Dramatic advances have been made at the interface between the nerves in the amputated limb and new bionic limbs. Sur-geons have actually rerouted arm nerves onto the chest of amputees so that each nerve can be directly mapped to sensors in the artificial arm, which is secured over the shoulder and with flaps onto the chest wall, providing the wearer with increasingly precise finger control and even a modified form of sensation.

The rate of evolution of cellphones and gaming devices is an obvious marker of the rapidly growing power and speed of microprocessors. Sensors, such as those used for artificial vision, artificial hearing and nerve signal detec-tion, are increasingly sophisticated. Better batteries provide longer life and more power in smaller packages. Wireless advances, such as the Bluetooth standard, have eliminated the need for awkward wired connections among components. These generic, enabling technological improvements in computing, sensing and battery power have been matched by specific medical innovations such as ever-better modeling of nerve signaling, surgical procedures that reroute nerves to improve neuroelectronic communications and the development of surgical implants specifically tailored to communications with particular nerves.

The cochlear implant for the deaf is a relatively mature example of a neuroelectronic device designed for people with a specific type of deafness in which the auditory or hearing nerve is intact and functioning but there is a problem with the mechanics of the ear. The implant consists of external and internal components. A microphone, speech processor and transmitter are worn externally like a hearing aid. The microphone picks up sounds from the environment; the speech processor filters the raw sounds and extracts bandwidths that are typical of human speech; and the transmitter sends the processed sounds through the skin to the internal receiver using electromagnetic induction. The internal, surgically implanted receiver/stimulator relays the sounds to an array of electrodes positioned along the auditory nerve.

With training and practice, a previously deaf individual can hear and understand speech in quiet environments after a cochlear implant. These devices are improving continuously, although at least one user described the quality of the voices he hears with his implant as robotic, analogous to "a croaking Dalek with laryngitis." As with the Audeo and Braingate, the key driver for the development of auditory technology is human disability, but it would be a trivial matter to take it one step further and tune the speech processor to hear wavelengths not typically available to normal human ears, such as the ultra- or subsonic frequencies audible to bats, dogs and elephants. It is merely a matter of time before bionic auditory implants will become available that have the ability to confer *supranormal* hearing capabilities on their recipients.

Optical prostheses are in a much more primitive state at present than their acoustic counterparts. There are two general approaches to the delivery of optical information to the brain in a way that bypasses damaged parts of the eye, such as the retina in macular degeneration and retinitis pigmentosa. Retinal neuroelectronic implants might be surgically placed in the eye to gather light projected onto the back of the eye through the pupil, and to then electronically relay it to the optic nerve. A more radical alternative might take information from an externally worn camera and relay it directly to the visual processing areas in the brain, thus bypassing the optic nerve altogether. Unlike cochlear implants, both approaches are considered experimental at this point, but improvements are likely to be very rapid over the next few years. As with auditory prostheses, optical prosthesis could readily be programmed or designed to recognize wavelengths outside the range of normal human perception. One day even normal people may have access to optical implants that can detect infrared and ultraviolet light.

While working as a power lineman, Jesse Sullivan lost both of his arms at the shoulder in an electrical accident in May 2001. Lucky to be alive, he was, however, profoundly handicapped. Today, thanks to a series of medical and engineering innovations, he has a functional bionic left arm (he uses a normal prosthetic arm on the right) that allows him to perform activities such as raking leaves, eating soup and shaking hands. Sullivan has undergone a series of nerve transfer operations that moved the nerves ordinarily used to control his left arm to locations across the left side of his chest. In a complementary effort, a prosthetic arm, known as the Boston Digital Arm, was developed by a Massachusetts company called Liberating Technologies. It can be strapped to the torso and is equipped with sensors mapped to lie over the rerouted nerves. Sullivan has had to relearn movements, much as a toddler learns, but he is able to perform a variety of complicated movements with the new arm. And the technology continues to improve: The latest version of the arm provides him with a sense of force feedback, the ability to sense how hard he is gripping an object.

Liberating Technologies is not the only pioneer in bionic prostheses. The U.S. Defense Advanced Research Projects Agency, which sponsored the development of the Internet, is currently coordinating the Revolutionizing Prosthetics Program, an effort to develop bionic extremities. Participants from Johns Hopkins and a variety of American and international partners are rapidly prototyping new prosthetic technologies including injectable muscle sensors. A Scottish company called i-LIMB has developed an artificial, multi-articulating hand with five independently powered digits including a rotating thumb and independent index finger (the three remaining digits operate as a unit). The hand can hold a piece of silverware, point with its index finger and grip a key or a glass. Bioprosthetic legs and feet have been developed to such a degree of sophistication that they are equipped with sensors to monitor speed and load on the extremity a thousand times a second and are coordinated by artificial intelligence logic.

The mental feats of Stephen Hawking are an almost unfathomable example of the potential and the will of the human mind, made all the more remarkable by the obstacles he overcame in communicating with the world in his lectures and books. Within the next decade or perhaps two, assistive technologies will mature to the point that they will actually become augmentative. Many of us will, at some point, be confronted with the option to enhance our existing toolbox of senses and appendages with higher-powered electromechanical alternatives, and there will be a range of ethical, financial and medical issues associated with these options.

Few of us are likely to have a problem with purchasing what we might call a *mental* mouse—the inevitable descendant of the Audeo. This device will dramatically increase the speed with which we control our computer; and a noninvasive neuroelectronic interface will be nonthreatening. At the other end of the spectrum, however, is what will happen when engineers develop the neurologically implanted descendant of the Braingate, providing supranormal input and output bandwidth over tomorrow's Internet or perhaps linked with powerful implanted auditory and visual prostheses?

While these surgically implanted devices will be initially designed for patients with various physical problems, the day will inevitably come when some features of neural implants are so desirable that healthy individuals will wish to undergo elective neural chip implantation. And if you think doctors will act as ethical gatekeepers and balk at elective brain surgery, I think you're wrong— given what we do in the way of elective plastic surgery today. Of course, as with plastic surgery, there will also be the inevitable issue of the haves and the have-nots. If Medicare is still around, it certainly won't be paying for elective neural chip implants; but you can be sure that the same folks who pay for surgery to look younger or better will pay a lot more to look farther or become smarter. And if those haves *truly* are smarter by some measure, or more connected, what, then, does that mean for the rest of us? Medicine will at that point have stepped beyond its edge.

The visionary science fiction writer William Gibson described the inherent possibilities of neural implants that can jack in and connect directly to the Matrix network in *Neuromancer;* the world he describes is very different from ours. Eventually, direct mind–computer communication with the Internet or some other matrix really will increase to the point that the world will undergo another currently unimaginable quantum transformation, comparable to what we experienced in the years between 1977, when I played Zork one letter at a time on an isolated mainframe computer through a primitive acoustic coupler, and today, when my children can play multiplayer, three-dimensional games like Halo on their personal laptops, across the Internet and with humans all over the world.

At the conclusion of the movie *A Brief History of Time,* Hawking's mother says: "He does believe very intensely in the almost infinite possibility of the human mind . . . I don't think that thought should be restricted at all. Why shouldn't you go on thinking about the unthinkable . . . Think how many things were unthinkable a century ago, and yet people have thought them. And often they seem quite unpractical. He's a searcher. People must think. They must go on thinking. They must try to extend the boundaries of knowledge."

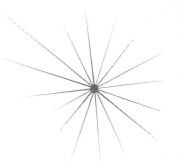

CHAPTER 7

WHICH ONE
WAS WHAT ONE?

In 1931, archaeologists working at a dig on *Playa de los Muertos* (Beach of the Dead) in Honduras unearthed part of the skull of what appeared to be a woman in her twenties with most of her teeth intact. As they examined the mandible, or jawbone, more closely, it became apparent that several of the incisors were actually made of shell. The teeth were made of shaped pieces of the shell of a bivalve mollusk, like a clam. The archaeologists initially assumed that the shells were implanted in the jawbone at some point after death, perhaps as a cosmetic procedure prior to burial, and left it at that. The skull was consigned to a museum shelf.

Forty years later, Amadeo Bobbio, a renowned Brazilian dental scholar, x-rayed the jaw and discovered that new bone had formed around the base of the shell teeth, indicating that they were placed during life and were essentially fully functional dental implants. This was remarkable because the dig was at a Mayan burial site from the sixth century A.D.

The fact that Mayan dentists were skilled enough to replace teeth is as astounding as another primitive medical procedure. We've known for a long time that trepanning, the intentional creation of a hole in the skull, was done by prehistoric cultures. It turns out, however, that while it's not clear whether the Mayans were performing restorative or cosmetic dentistry when they implanted the shell incisors, they were very skilled in shaping and decorating teeth. They filed and inlaid incisors (the visible front teeth) with jewels for ritualistic or religious reasons. And Mayan dentists used tools that are in many ways as functional as the ones we use currently in dental offices, including precise drills so that they could insert a variety of precious stones into predrilled cavities in the teeth.

Front teeth were also filed into geometric patterns or shapes, and dentistry obviously played a big role in the Mayan concept of beauty and, perhaps, wealth. One can imagine the sense of anticipation mixed with apprehension that the young Mayan woman must have felt before her dental procedure. I'm sure her emotions were very similar to those experienced by modern men and women, now so well chronicled on reality television shows, as they prepare for cosmetic surgery, body piercing or tattooing.

There are a variety of ways in which cosmetic dental or bodily surgery, body piercing and tattoos are similar, aside from the fact that they all serve some often-fleeting concept of human beauty. Each of these procedures, be it a breast augmentation, ear piercing or skin art, has potential medical complications that range from infections, both bacterial and viral, to immune reactions, whereby the body rejects the foreign material.

One of the most basic requirements for a living organism is the ability to distinguish the self from the non-self, and to eliminate the latter. This would have been a straightforward problem for the first unicellular organisms, because the self was simply everything within the cell's membrane—everything else was non-self. Life became much more complicated during evolution and the development of complex multicellular organisms, in which a variety of cells had to coexist and cooperate in a single coherent entity. Such an organism needed to develop a method of saying "all of these cells belong to me and anything else is foreign." Evolution then designed a host of creative methods of killing, and sometimes eating, the non-self.

Theodore Geisel, better known as Dr. Seuss, described this issue very succinctly in "The Sneetches," a 1960 poem tome about two races of creatures who lived, like our Mayan girl, on a beach. The Plain Belly Sneetches are distinguished from Star Belly Sneetches by the presence or absence of a visible star on their pot-bellies until a "fix-it-up chappie" named Sylvester McMonkey McBean shows up with a machine that can tattoo stars on plain pot-bellies and remove them from Star Bellies with equal facility. By the time Sylvester drives his Star on/Star off machine away from the beach, the Sneetches are penniless and McBean is rich. The Sneetches, furthermore, are no longer able to distinguish who originally belonged to which group, which is exactly the goal of modern immunologists and transplant doctors who would like to be able to introduce foreign tissue or insert foreign materials into a recipient without provoking an immune reaction.

The immune system recognizes things as foreign based on their surface characteristics, so white cells and antibodies are constantly sniffing about looking for non-self surfaces, be they animal, mineral or vegetable. When scout cells encounter something that they don't like the look of, they summon in more white cells and that area of the body becomes inflamed, both figuratively and literally. First-responder cells recruit other types of white cells, blood vessels become leaky and the area swells as the body attempts to wall off the non-self substance and prevent it from spreading further. Typically, the foreign body must be removed or the immune system suppressed before the inflammation subsides.

In the spring of 1990, Janet Bowen was an unemployed mother of two, who had delivered her second and, what she and her husband had decided was their last, child, a little boy. Bowen had been an active, attractive, fit twenty-year-old mother when she married, a recreational runner who had completed a half-marathon at one point. She was now ten years older and twenty pounds heavier, and she had the self-perception to realize that she was at one of those branch points that come up periodically in all of our lives.

She felt like she could, on the one hand, go the route of many of her friends and resign herself to motherhood, putting her own appearance and self-esteem aside for a bit while she concentrated on raising her kids. On the other hand, she believed that with a little work, she could get back to being and feeling attractive and proud of her body. She decided to go with Plan B and set about it methodically.

With her husband's support, she hired a babysitter several days a week and joined a health club, where she worked out religiously, doing a combination of the latest aerobic fad and some weights. She was an enthusiastic practitioner of step classes, and later Tai-Bo, spinning and power yoga. Eventually she hired a personal trainer and she definitely did get more fit, but she was never able to lose what she described as the "saddlebags" on the back of her upper thighs. After seeing a lot about liposuction on television and in magazines, she concluded that was what she needed.

After a little research, Bowen found a local dermatologist who "felt that I was a perfect candidate, because I wasn't obese and I had a specific area that could be addressed." He advised tumescent liposuction, which is also called lipoplasty, lipectomy or, by artistically inclined, market-savvy surgeons, liposculpture, in which a combination of salt water, an anesthetic and epinephrine are injected into fat, which is then suctioned from beneath the skin.

Bowen's liposuction went smoothly and after a few weeks she felt rejuvenated, at least in regard to her nether half. This initial, successful foray into cosmetic surgery, a field about which she had previously felt some scorn, compelled her to think about a second, in some ways more visible, problem area—her breasts.

Although she had never really obsessed about her breasts before having kids, the pregnancies had definitely changed their shape for, what she considered to be, the worst, and she had come to rely on the various artifices used by bra and bathing suit manufacturers to feel comfortable wearing a low dress or going to the beach. She floated the idea of breast augmentation with her husband, who was enthusiastically supportive, and suggested that she start looking for a good plastic surgeon immediately.

Bowen did what many patients do at that point, and started her research on the Internet, where she found hundreds of glossy Web advertisements for plastic surgeons within a couple of hours' drive. She learned about sizes, different incisions, subglandular, subpectoral and submuscular placement, textured versus nontextured implants and silicone- versus saline-filled products. She also learned about complications, although most of these were given lesser play on the surgeons' Web sites than they were on those of medical schools and professional societies. She joked with her husband about the fact that men too can get breast implants to "create a larger, more defined chest."

At the time, in the early 1990s silicone implants were preferred because they were believed to provide a more natural look and feel, and Bowen decided to go forward with implantation of C-sized, silicone gel implants. On the scheduled day of surgery, she changed her mind at the last minute and told the surgeon she wanted to change to D cups. Several weeks later everything looked great—the swelling and pain had disappeared, and you could barely see the scars.

After a year, however, the natural feel of her breasts had disappeared. She developed hard, painful areas (she was told they were "capsules") around both implants, and underwent a series of "closed capsulotomies," in which her surgeon essentially squeezed each breast as hard as possible to break up the scar tissue that reformed in each intervening period. It was very painful, but he advised her to ask her husband to do the same thing at least weekly. Once formed, these capsules often contract painfully around the implant, and at one point she actually had to have surgery on her left breast, during which the implant was removed, the scar tissue reduced, and the implant replaced.

Coincident with or perhaps due to her breast problems, Bowen's health had deteriorated noticeably. She stopped exercising because she had developed more or less constant muscle pain. Her surgeon reassured her that, despite the U.S. Food and Drug Administration's (FDA) moratorium on silicone implants, there was no conclusive evidence linking them to disease. At the suggestion of her regular internist, Bowen finally had a mammogram. It showed that one of her implants had ruptured and that there was "silicone extravasation." She had read of this in the discussions on the breast implant support groups she found on the Internet. She understood that the silicone was now able to freely move throughout her body and knew that was not a good thing.

The lymph nodes of the armpits are the first line of defense for infections in the breasts and the first location to which the syrupy, extravasated silicone migrates. The human immune system generally tries to wall off the silicone in the nodes by the formation of scars and lumps of immune tissue called granulomas, which can be mistaken for cancer. Eventually, silicone can spread to other areas, including the lungs. Although the FDA couldn't prove an association between ruptured implants and cancer or autoimmune diseases, they did note, in a large study reported in 2001, that there was a higher incidence of fatigue and muscle pain symptoms in patients who had proven ruptures.

Bowen eventually decided that she needed to have the implants removed. She had this done by a different plastic surgeon because she had become uncomfortable with the original doctor. After a period of several months, her energy began to return and she stopped having muscle pain. She thought about the possibility of replacing the silicone implants with newer saline-filled versions but, in the end, after talking it over with her husband, decided not to.

Silicone implants have also been designed to augment calf, bicep and tricep muscles as well as the buttocks. Silicone is not the first foreign substance used to enhance human appearance that has caused problems. Contact dermatitis is an example of an immune response on the skin, most commonly resulting from exposure to the resin of poison ivy. However, the second most common cause is an allergic reaction to nickel, a base metal frequently used in costume jewelry (and less frequently found in nickel coins, which are mostly copper).

Patients with nickel dermatitis can't wear, or in some cases even handle, alloys containing the metal, which include some types of lower-quality silver and gold, and therefore often don't wear jewelry. There are reports in the medical literature of bank clerks developing eczema from repeated contact with nickel-alloy coins, and one woman who developed dermatitis after storing nickels in her brassiere. Cobalt, mercury, chrome and even purer gold also provoke allergic reactions in some people: Gold and mercury amalgams, for example, can cause inflammation or ulcers of the mouth when used in dental work.

Gold has been used for dental prosthetic work for a very long time. While the Mayan mandible is the first known example of a successful dental implant of any sort, Egyptians and Etruscans used gold to create fixed partial dentures or bridges. The Egyptian pharaoh in about 3000 B.C. had a "Chief Toothist" named Hesi-Re. Both cultures used gold to secure a replacement tooth to adjacent teeth.

Gold has also been used in combination with silver, porcelain and glass to create prosthetic eyeballs. The earliest prosthetic eye, discovered in an area along the border between Iran and Afghanistan, was made of bitumen paste with a golden iris inscribed on the surface. Other prosthetic eyes made of precious stones and jewels have been found in Egyptian royal tombs. Imagine how awe-inspiring it must have been to be subjected to the seemingly godlike gaze of a ruler with a golden or bejeweled eye.

Captain Ahab's prosthetic leg, replacing the one nipped off by his nemesis in Herman Melville's *Moby-Dick,* was fittingly fashioned of ivory, a durable and decorative material, and it was probably modelled after real-life ivory prostheses, not just some fanciful creation of the author. It may have given the troubled captain some sense of retribution to be stumping around on the tooth of a dead whale, albeit not the one he would have most liked to terminate. Not surprisingly, Ahab had an uneasy relationship with his ivory extremity and was found at one point prone and insensible on the streets of Nantucket, from where he eventually sailed on the *Pequod*, with his peg leg having "stake-wise smitten, and all but pierced his groin."

Much as ivory was a good material with which to make limb prostheses because it was durable and not too heavy, gold was an ideal material for early "toothists" to work. It is a strong yet malleable metal, doesn't corrode, causes minimal immune reaction and almost certainly provided some decorative and perhaps status statement. Gold has largely been replaced in the field of cosmetic dentistry because the whole field of medical bioengineering and materials is in the midst of a revolution that is changing the way we develop and manufacture human implants.

NASA's underwater research environment in the Florida Keys—the NASA Extreme Environment Mission Operation, or NEEMO—is named in deference to Jules Verne's mysterious fictional explorer Captain Nemo, who made his undersea voyages in a submarine called the *Nautilus*. Nemo and Ahab share the distinction of being two of the most compelling, and peculiar, sea captains in literature; and both ran into the elusive giant squid during their voyages. The *Nautilus* attacked one of these massive tentacle denizens of the deep, and another makes an ominous appearance in *Moby Dick:* seeing a squid surface near his whaling skiff, Ahab's first mate, Frank Starbuck, exclaims, "The great live squid, which, they say, few whale-ships ever beheld and returned to their ports to tell of it." To which Ahab, perhaps prophetically, had no response.

The giant squid has always been an intrinsically mysterious and seldom-seen denizen of the deep; the first pictures of these sometimes 45-foot-long cephalopods weren't taken until 2004. They have also perplexed bioengineers who, until recently, weren't able to figure out how the squid keeps from chewing itself apart because its big sharp beak attaches to a largely gelatinous body. Paraphrasing one scientist's analogy, imagine what would happen if your hand were made of Jell-O, and you set about carving up a fish for dinner with a handleless knife. The knife would, of course, slice right through your fingers. Yet giant squid chomp away on deep-sea fish like roughy and hokie that they spot with their dinner-plate-sized eyes, and fork them into their razor-sharp, parrotlike beaks with their floppy tentacles.

It turns out that Mother Nature found an elegant method for marrying the hard beak to the soft body. The beak is made of *chitin,* the same material in crabs' hard shells; and the beak changes density gradually from its tip to the muscular attachment at the mouth. The chitin at the tip is a hundred times harder than that at the base, and the degree of softness is a function of how much water is mixed with the chitin at any point along the beak. While these massive creatures of the deep are bigger and stronger than almost anything else they encounter, their main predators are leviathans like the Great White Whale. Sperm whales are often found with beak and tentacle scars on their head, which gives us some hint, some shadow-play of the life-and-death drama that must go on in the dark, cold deeps.

Bioengineers who design materials for implantation in the human body—artificial teeth, breast implants, joint prostheses or heart valves—are increasingly attentive to and sophisticated about the interface between the self and non-self. The materials they select for various applications are based on their color, strength, cost, resistance to wear and biocompatibility—meaning the degree to which they may provoke an immune response. They have learned lessons from the successes and failure of earlier implants, such as the silicone-gel breast implant and early heart valves. They adopt design features from Nature. For example, the graded interface between soft and hard materials in the squid beak is directly relevant to the problem of how to attach a hard mechanical heart valve to a soft aorta or how to create a new joint surface to replace the deteriorated original.

Over the last fifty years, the development of biomedical materials for medical care has become both a science and an industry. Specific substances have been designed to repair or replace damaged, missing, diseased or worn-out organs, bones, joints and limbs. The range of materials we use is extraordinary, from metals to ceramics and polymers; from animal parts to human parts salvaged from our dead; from cellular scaffolding implanted in the body to human structures and, perhaps in the future, organs custom-grown for implantation back into the body from which that organ's cells were originally harvested.

Earlier in this chapter I mentioned the problems that the first living organisms faced in the transition from single cells to multicellular organisms. At the point of differentiation, the organism was forced to evolve a method of distinguishing the self from the non-self, and a way to eliminate the latter. The immune system developed as a sort of cellular police, and the ability to recognize and repel foreign material is common to all higher life forms.

Another requirement of a multicellular organism is a repair mechanism. When a single-cell organism dies, that's pretty much the end of it. On the other hand, if you are a multicellular organism and one of your specialty cells dies, it is extremely useful to have a way to replace it, lest your whole organism becomes less fit and therefore vulnerable. Like immunity, the capacity for self-repair is another fundamental property of life, extending even to the level of DNA. Bioengineering is the medical discipline that seeks to restore or improve function when body repair mechanisms fail by, for example, joint replacement, while avoiding engagement with the immune system.

There's another issue confronting bioengineers related to the fact that life evolved from sea-water. The first multicellular organisms that tentatively scrambled on to the beach were probably mostly salt water with fragile membranes and no skeleton—like jellyfish. We humans, while quite a bit down the evolutionary trail, are still about 70 percent salt water, albeit somewhat more dilute than sea-water. As you'll know from any trip to the ocean, salt water is extremely corrosive, posing one of the many challenges a bioengineer faces in designing materials for implantation in the human body.

Many of us will remember a time in our youth when everything about our bodies worked perfectly: We could see without glasses, our hair was all there and its original color, our muscles and joints never ached, sleep came easily and everything worked just as nature apparently had designed. Many of us will also remember that bittersweet point at which some part of our once maintenance-free body changed or broke *permanently*. For some it was the need for vision correction, for others the loss of a tooth, for yet others a newly dysfunctional joint. Fortunately, in many cases bioengineers have designed pretty good medical workarounds that limit the impact of bodily deterioration. But, the body's capacity for self-repair is not limitless and the replacement parts are not nearly as good, in many cases, as the factory-equipped version.

The mammalian heart valve is an example of Mother Nature's best work and we're born with four of them, all slightly different. The heart beats about four thousand times an hour, and as many as three billion times in an average life-span; and it seems almost inconceivable that they hold up that long if you've ever had the opportunity to look at one of these seemingly flimsy pieces of tissue.

The aortic and mitral valves are on the left side of the heart and they're very different in appearance. The mitral valve, for example, has two leaflets and opens and closes like a set of smiling lips; while the aortic valve, viewed from the top down, is shaped like the Mercedes-Benz logo, with three leaflets, each of which is shaped like a little scallop shell. As it opens and blood is propelled out of the heart, the lacy tissue that forms the valve leaflets folds up against the wall of the aorta into a slightly recessed area called the Sinus of Valsalva, named after Antonio Maria Valsalva, a prolific Bolognese anatomist of the seventeenth century. Both the mitral and aortic valves are extremely efficient at letting blood flow through in one direction, with essentially no restriction when they are open, and preventing backflow when closed. Nevertheless, valves are subjected to tremendous stresses as they open and close, from both the turbulence of the blood flowing by and the shearing mechanical stresses of opening and closing. The history of the development of artificial heart valves is dramatic and shows brilliant bioengineers and physicians at work.

As with many medical advances, the origins of heart surgery lie in a time of war—World War II in this case, in which shrapnel and bullet wounds to the heart were common. A young Army surgeon, Dr. Dwight Harken, worked with animals to develop methods to remove metal from the beating heart. Once he was able to demonstrate consistent success in an animal model, he moved on to humans, and he eventually saved the lives of many young soldiers. Harken, working in Boston, and Dr. Charles Bailey in Philadelphia were early pioneers of another innovative procedure that allowed surgeons to repair the mitral valve while the heart was still beating, a seemingly impossible feat of legerdemain. The surgeons essentially picked the pocket of the organ, making a small incision in the atrial chamber, putting a purse-string suture around it, slipping a finger into the still-beating heart with the purse string pulled tight to prevent leakage and cutting open the defective mitral valve with a special knife or scissors designed for this purpose.

This operation was typically performed to fix valves damaged by rheumatic fever; as many as half of the original patients undergoing what is called closed mitral commisurotomy died on the operating room table, although surviving patients often had dramatic improvements. Harken and Bailey had an intense rivalry and, according to Anthony Dobell in "Rival Trailblazers," a 1996 article in *Annals of Thoracic Surgery*, "each criticized the other's work . . . they quibbled about nomenclature, argued about the other's interpretations, disagreed on clinical assessment and mortality figures. In fact, they rarely agreed on anything, and when they did, they tended to congratulate the other for finally coming around to a proper understanding." Both surgeons persevered in the face of initially very poor results. Bailey actually operated on two patients with the same initials and same disease on the same day in 1948, and drove across Philadelphia from one hospital to another to do so. Both patients died on the table.

I can assure you that there is no more emotionally wrenching event in a physician's life than to have a patient entrust himself to you in the operating room and then watch him die in your care. This happened to me recently with a very sick, very frightened elderly woman, whom I attempted to reassure before surgery saying "we'll take very good care of you." We did, but she died. Bailey, Harken and most of the great surgical pioneers faced this outcome frequently early in their careers. Dr. John Gibbon, who invented the heart–lung bypass machine, stopped performing cardiac surgery after two early, tragic surgical deaths in young children. The development of the artificial heart valve was an exemplary bioengineering challenge that took great courage to press on to success.

The first artificial heart valve used what's called a ball-and-cage design originally patented as a "bottle closure invention" in 1939 by Edward Stephany. As the pourer inverted a bottle, the liquid within would push the ball into the cage allowing flow; and when it was then set upright, the ball would fall back onto the bottle's mouth and reseal it. As Stephany explained in his 1938 U.S. patent application (patent number 2,177,310), the design "is particularly useful at beverage bars for facilitating the pouring of beverages without removing the stopper and protecting the liquid against insects and other foreign matter, as well as evaporation." The caged-ball design was adopted by a cardiac surgeon–hydraulic engineer team, Dr. Albert Starr and Lowell Edwards, to create the first artificial heart valve. Before inserting the valve in a human, however, Starr and Edwards had to confront problems characteristically faced by bioengineers in the design of materials for implantation in the human body.

The first artificial valves implanted in dogs tended to tear away from the aorta along the sewing ring, due to a mismatch between the stiffness of the valve ring and the much more flexible aortic tissue—the same problem faced by prosthetic joint manufacturers when an artificial joint is stiffer than the bone into which it is inserted, and the same problem nature solved in the evolution of the squid's beak. The force of a heartbeat or a footstep becomes concentrated at these points of mismatched flexibility at the interface of the human body and the implant.

Starr and Edwards solved their problem by cushioning the sewing rings with cloth to buffer the aorta from the jarring effects of the mechanical valve opening and closing. This is a poor man's version of the more elegant solution of the continuous gradation of flexibility along the chitin in the squid.

A second problem with the ball-and-cage valve design was the formation of clots on the prosthetic valve. Starr and Edwards believed, correctly as it turned out, that the back and forth movement of the ball in the cage would reduce clot formation. To be fair, all of their patients required life-long blood thinning, or anticoagulation, using a synthetic drug called warfarin (which was originally used as a rat poison). Warfarin is chemically related to coumarin, a naturally occurring chemical in many plants that was originally discovered during an investigation of cattle that had spontaneously bled to death after eating moldy clover.

The Starr-Edwards valve represented an extraordinary advance in heart surgery. Starr's first human patient lived for ten years after replacement of his mitral valve and eventually died after falling off a ladder. The original cage was made of Lucite, also known as Plexiglas, a durable plastic, and the ball was made of silicone rubber. Starr's successful valve replacement started a wave of innovations in heart surgery all over the world, eventually leading to the development of a tissue heart valve by Alain Carpentier, a French surgeon, and to the first heart transplant by Dr. Christian Barnard of South Africa. The caged-ball design has subsequently been superseded by toilet-lidlike valves, tilting-disk valves and, more recently, bileaflet valves whose two semicircular leaflets open and close like swinging doors.

The newer valve designs have several advantages over the Starr-Edwards valve. Blood flowing through the caged-ball valve is very turbulent and red blood cells fragment, causing the development of anemia; this is not a problem with newer valves. Modern valves are less likely to provoke clot formation, and patients can therefore be treated with much lower doses of anticoagulants. The leaflets are made of pyrolytic carbon, which is light, resistant to clot formation and very durable. In fact, this type of carbon is so hard that it is used to encapsulate nuclear fuel rods and missile nose-cones. Newer valves are also much more efficient: A bileaflet mechanical valve allows blood to flow with much less resistance than a caged-ball valve of the same diameter. Interestingly, these valves are modeled on double-doored devices used in irrigation canals in India. However, in spite of the improvements in mechanical valves, surgeons would prefer to use a model as close to the original as possible.

Dr. Carpentier, who still lives in France, developed the tissue heart valve after one of his patients had a large stroke due to the formation of a clot on a Starr-Edwards valve. The patient was an artist and became significantly disabled after the stroke; Carpentier felt that a tissue valve would be much less susceptible to clot formation and would prevent this debilitating outcome. The first valves he used were taken from human cadavers, but they tended to become infected because of inadequate post-mortem preservation. Carpentier turned to valves harvested from pigs and coated in a mercury solution to sterilize and preserve the animal tissue. Unfortunately, the mercury provoked an intense immune response, which caused rapid deterioration of the valve.

Today's tissue prostheses are valves in which pig, cow or horse pericardium—the sack that encases the heart—is formed into leaflets and draped over a metal skeleton. These tissue valves are processed in such a way that all of the animal surface features that might activate the human immune system to respond to the non-self are removed.

The progressive evolution of artificial heart valves exemplifies many of the issues confronted by physicians and engineers in designing replacement parts for the human body. The materials used in the various heart valves I've described include plastics, rubber, metal, crystallized carbon, ceramics, polyesters, animal tissue and human tissue. These same materials are used in various configurations and combinations to replace joints, substitute for bone in the middle ear, repair damaged blood vessels and create new organs. The totally artificial heart, for example, is entirely constructed of manmade materials.

Bioengineering innovators are as creative as any engine tinkerer—witness the adaptation of a bottle closure design to a heart valve. But while the first pioneers, groping around for solutions, used materials like mercury that seemed good but weren't, bioengineering science has now progressed to the point that bioengineers manufacture new materials fully cognizant of the potential roles that stress, friction, corrosion, immune response clotting and other body-biomaterial interactions will play in the success or failure of the new part.

Artificial organs have been designed to replace most of our own. The artificial kidney, or dialysis machine, is arguably the most successful and certainly the oldest, dating back to the 1940s. Many of us also remember the Jarvik artificial heart implanted in Barney Clark, a 61-year-old retired dentist, in 1982. Clark survived 112 days but was tethered thereafter to a drive mechanism the size of a refrigerator, and eventually died following a blood clot to the brain.

While Dr. Robert Jarvik is popularly credited with the invention of the artificial heart, his design was largely based on patents assigned to the University of Utah by a ventriloquist, dancer, actor and inventor named Paul Winchell. Winchell's roles included Fleegle in the American television series *The Banana Splits* and Tigger in the Disney movie *Winnie the Pooh*. Clark's heart was a clunky device made of plastic, polyurethane and metal; today's totally implantable artificial heart weighs two pounds, has internal surfaces that are resistant to clot formation and is made of titanium and a proprietary polyurethane plastic that inhibits clot formation.

No artificial lungs or livers have yet been designed for general use, but solutions are on the horizon that could be used as a bridge to transplantation, much as temporary heart support devices do now. The pancreas can also be transplantable, but a prototype biocomposite pancreas sequesters living pancreas cells inside of a silicon and titanium chamber etched with pores that are large enough to allow the passage of glucose, insulin and oxygen, but too small to let immune components by. A device like this may soon be implanted as a permanent artificial alternative to a real pancreas.

Advances in laser technology have enabled us to sculpt the human lens for better *visual* acuity, and bioengineers have developed implantable, artificial lenses made of biologically inert polymers. However, the approaches used to replace *skin* and restore *hearing* give us the best clues about what bioengineering will look like in the future.

Conducting hearing loss results from damage to one or more of the three bones of the middle ear, collectively known as ossicles or little bones. The bones conduct sound impulses from the eardrum to the inner ear, where they are translated into the nerve signals that are subsequently processed by the brain. The scientific names of the bones are the malleus, the incus and the stapes, taken from the Latin words for hammer, anvil and stirrup. When the eardrum vibrates, it moves the hammer, which acts with the anvil, like a lever, to move the stirrup. The middle-ear bones translate air-based sound waves into a fluid wave in the inner ear, dampening excessively loud sounds when necessary to protect the inner ear.

The materials that bioengineers and surgeons originally used to replace damaged ossicles included titanium, Teflon, platinum and plastics. Today's replacement ossicles, however, have been engineered to more closely adhere to natural design principles, and are made of composites of plastic or glass with biologically active materials that stimulate bone growth. After implantation, the new ossicles become covered with new bone and eventually become superficially indistinguishable from the original ones. The first words heard by the recipient of one of these new prototype middle ears were "Hamburger, hot dog, ice cream."

Biocomposite materials that integrate natural and man-made features, similar to those used in the ear, are also being used to manufacture new skin for badly burned patients who lack enough of their own that's suitable for grafting. Collagen, one of the body's natural connective tissues, is bound to silicone in a two-layered sheet. This artificial skin material is applied to a burned area with the silicone on top; new skin eventually grows into the collagen over a period of weeks while the silicone protects the vulnerable area. The silicone is eventually peeled off when the new skin is mature. The collagen underlayer serves as scaffolding through and over which the new skin grows, like ivy on a trellis.

Huge advances have been made in the development of inert materials that can duplicate or replace functions in the human body without triggering the immune system, but the holy grail for bioengineers is organ engineering, or the de novo construction of living, functional organs and components that can regenerate or replace damaged tissues in the body. In order to be successful, organ engineering will require the isolation, proliferation and differentiation of various component stem cells, and the design of scaffolds, or framing, to coordinate the growth of three-dimensional tissue organ structures.

In effect, organ engineers will need to function as architects. Their organ buildings must be structurally sound; a variety of scaffolding materials have become available, some of which are permanent, such as nanofibers or porous textiles, while others are resorbable or biodegradable. Chitin, for example, is the principal component in crab's shells and squid beaks, and yet it is also used to make resorbable surgical sutures. The bricks and mortar of the organ are its constituent cells, which may come from mature cells grown in cultures, like crops, or may proliferate from precursor cells, like stem cells, which we'll discuss in more detail later. Finally, the bioengineered organ must be designed with provision for utilities—in this case, blood and lymphatic vessels, rather than plumbing and HVAC.

Intuitively, it seems almost inconceivable that human engineers would be able to re-create, say, a kidney, which is a complex organ with a complicated and highly sophisticated internal architecture of arteries, arterioles, capillaries, venules, veins, urinary conduits and lymphatics. A neokidney of the future, however, doesn't have to *look like* a kidney, which is admittedly odd looking. There's no anatomic reason that the neokidney couldn't be shaped to look like the corporate logo of its manufacturer. And while the artificial assembly of an organ of such complexity would certainly be challenging, the challenges are no greater than those involved in the design and assembly of modern computer microchips. In fact, some of the techniques used in computer chip design are just what will be required for organ engineering. Computer assisted design and manufacturing (CAD/CAM) can be used to create an appropriate three-dimensional organ scaffold that can then be populated with cells, using techniques very similar to ink-jet printing technology.

We humans have been endlessly inventive in finding ways to use natural materials such as shell fragments, to alter our own cells and tissues, and, increasingly, to design new materials like pyrolytic carbon in the service of human appearance and health. Advances in bioengineering, miniaturization, nanotechnology and stem cells are converging, and we are on the verge of an era in which we'll be able to repair or replace injured or aging organs, tissues and joints. Manmade prosthetic materials will eventually transition seamlessly into our own tissues—exactly as the squid's beak gradually hardens to the tip—turning today's cosmetic, orthopedic and cardiovascular implants into antiquated museum artifacts.

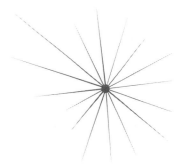

CHAPTER 8

LITTLE MINDS

"**N**o more pleasant sight has met my eye than this of so many thousands of living creatures in one small drop of water, all huddling and moving, but each creature having its own motion . . ." Anton van Leeuwenhoek wrote these lines in 1678 after examining a drop of pepper-infused water under one of his homemade microscopes; the creatures were, of course, bacteria, which he described, in a letter dated 1674 to the secretary of the British Royal Society, as "animalcules." It takes a leap of faith to believe in that which one cannot see; and whereas the members of the Royal Society had previously been most pleased to publish Leeuwenhoek's observations of *inert* microscopic objects, they balked at the idea that there might be heretofore unimagined forms of *life*. They initially refused to publish his descriptions of these microscopic organisms.

Eventually, with the help of lawyers, physicians and even a vicar whom Leeuwenhoek enlisted as witnesses, the group saw the light and the animalcules. In the end, they honored him with fellowship in the Royal Society. The former city hall janitor in the Belgian city of Delft had come quite far. Over the succeeding centuries, many medical advances have resulted from our ability to examine and manipulate the microscopic world; and we'll see in the next few chapters how medicine in the miniature will affect our future.

It would be nice to think that the disbelief with which the Royal Society greeted Leeuwenhoek's description of microscopic life was a product of the times, but consider the story of Stanley Prusiner, who discovered some *really small,* really *bad* actors—for lack of a better term—that can be just as deadly as bacteria but blur the distinction between the quick and the dead. At the time, it was widely known that something bad was happening to the Fore aboriginal natives of Papua, New Guinea. In the mid-twentieth century, they developed an inevitably fatal syndrome called kuru—characterized by headaches, joint pains and trembling—in epidemic numbers. When it became apparent that the Fore engaged in cannibalism, eating the bodies of the recently dead, a food-borne transmissible agent was suspected. This hypothesis was supported by the fact that kuru was much more prevalent among the female, child and elderly tribal members. The women primarily prepared and ate the brain, feeding what they may have perceived as tasty morsels to their children and parents.

Kuru shares pathological features with a number of other human and animal brain diseases, collectively called spongiform (spongelike) encephalopathies (diseases of the brain). Some of them are transmissible from one human to another, like kuru, while others, such as fatal familial insomnia, are clearly inherited. There are also a number of animal variants, the best known of which are mad cow disease and scrapie. The primary symptoms of all of these diseases are those of brain dysfunction, including cognitive, memory and movement problems, eventually progressing to death. There is still no effective treatment.

The cause of the spongiform encephalopathies was unknown when Prusiner started his research, but was almost universally believed to be a virus of some sort. All viruses carry a length of genetic code in the form of nucleic acid, with which they reprogram infected cells to make more viruses. This is directly analogous to computer viruses, which carry a length of computer code with which they reprogram *their* hosts to make more viruses. Prusiner spent a long time looking for a causative virus early in his career and eventually concluded that the infectious agent was a previously unknown infective agent: a defective variant of a normal cellular protein. He announced the discovery of what he described as "prions" or proteinaceous infectious particles in 1982. He proposed that prions are formed when an error in the nucleic acid code responsible for the formation of a *normal* protein results in the production of an *abnormal* version of the same protein. The abnormal protein, or prion, acts as a seed that converts normal protein to prion variants. And, because prions are not degraded by normal cellular housekeeping enzymes, the infected brain cell eventually bloats and dies—creating the hole in the sponge, as it were.

Prusiner's announcement was met with what he described as "a firestorm" of criticism, much as I imagine Leeuwenhoek's description of living creatures must have been by the Royal Society. Prusiner said: "the attacks of the naysayers at times became very vicious." The virologists who had staked their careers on the discovery of the causative organism were not best pleased. As Ralph Waldo Emerson put it in his essay "Self-Reliance," "a foolish consistency is the hobgoblin of little minds, adored by little statesmen and philosophers and divines."

I've heard Prusiner speak about his research and the initial reaction to it on a couple of occasions, and he is still, 25 years later, clearly very much in touch with the emotions of what must then have been a very painful period. As he elaborated: "The press was usually unable to understand the scientific arguments and they are usually keen to write about any controversy . . . While such scorn caused Sandy [his wife] considerable distress, she and my two daughters, Helen and Leah, provided a loving and warm respite from the torrent of criticism that the prion hypothesis engendered." Prusiner won the Nobel Prize for his work in 1997, and the forgoing quotations come from his autobiographical statement written at the time of the award and published in *The Nobel Prizes 1997*.

In many ways, medical progress over the past quarter century has been like Alice's long tumble "down, down, down" the rabbit hole in Wonderland—ever closer but never quite to the center of things. Like Alice, really great investigators just get "curiouser and curiouser," as they go through their careers, regardless of the adversity they meet with. And nowadays it seems as if the first thing one must do before setting off on a career in medical research is to follow the directions on the "Drink Me" bottle, the contents of which shrunk Alice, to be able to work with things really small. Prions, for example, are so small they are visible only under electron microscopes because the disease-causing proteins are really just big molecules.

Just as there has been an inexorable trend toward investigation of smaller and smaller things, like prions and the nucleic acids that make up our genome, there has been a parallel trend toward miniaturization of medical diagnostic devices and tools. This is no coincidence because the technologies used to ex-amine small things are tightly linked to the technologies needed to make small things. The evolution of the artificial heart pacemaker typifies this trend.

The first artificial, implantable cardiac pacemaker was designed by Rune Elmquist in 1958, who was working at the time with Åke Senning, a cardiac surgeon at the Karolinska University Hospital in Sweden. Elmquist was trained as a physician but spent his career as an engineer and inventor, eventu-ally forming a company that became part of what we now know as Siemens. They were working on animals to develop the pacemaker, and their research was described in lay publications, where it was discovered by Else-Marie Lars-son. Arne Larsson, her husband, had contracted hepatitis and, as a result, had developed inflammation of the heart, or myocarditis.

One of the potential complications of myocarditis is a syndrome called third-degree heart block, in which the "electrical wiring" of the heart is dam-aged, causing slowing or interruptions in the heartbeat. Arne Larsson's heart-beat occasionally dropped to as low as 28 beats per minute, which, while good enough for a trained athlete, caused fainting spells in the 43-year-old Swede. Else-Marie and hospital aides were occasionally forced to thump on Arne's chest 20 or 30 times a day to keep him alive.

The only pacemakers available at the time were large machines that were essentially stationary, much like some of today's heart-assist devices. The Larssons were told that Arne would probably die at some point from a pro-longed heart stoppage. Mrs. Larsson didn't accept this, so she contacted Elmquist and Senning and pleaded with them to make an experimental pace-maker suitable for implantation in a human.

The two researchers eventually agreed and designed an experimental, fully implantable device. Partially constructed in Elmquist's kitchen and the size of a hockey puck, it had only two transistors and required batteries that had to be recharged every few hours. This first implantable human pacemaker was in-serted in Arne Larsson on October 8, 1958, in a multihour major operation during which his chest was opened. The pacemaker failed in the first three hours, but fortunately there was a backup that lasted the next three years.

Arne Larsson went on to have 25 pacemaker placements after that first operation and lived to the age of 86, when he died of complications of malignant melanoma. Each pacemaker he received was smaller and more sophisticated than the last. Current pacemakers are the size of a half dollar, have 500,000 transistors and batteries that last years. Pacemaker placement is now performed as a brief procedure that does not require general anesthesia.

Larsson was a highly regarded, world-traveled man at the time of his death and had received many awards for his professional work as an engineer and his humanitarian work in patient advocacy. Despite having developed a fatal disease at the age of 43, Larsson went on to outlive both Elmquist and Senning. Two professors from the Karolinska University Hospital wrote these words about Larsson, Senning and Elmquist in an obituary that appeared in the *Journal of Pacing and Clinical Electrophysiology* in 2002: "[They] became great examples of how a combination of a brave patient and a bold physician and a creative engineer may change the world for numerous patients."

The variety of technical issues that plagued the first generation of pacemakers included battery failure, runaway pacing, broken wires and leakage of body fluids into the pacemaker housing. The corrosive effects of the salt water that circulates through our bloodstream wreaked havoc on the early pacemaker boxes. Today's versions use metals developed for the aerospace industry. Although the technology has improved dramatically, pacemaker problems continue to the present. For example, a search of the Internet for the terms "pacemaker" and "recall" returns a long series of Web pages from law firms offering to assist you, should you or a loved one be injured by a defective device, and offering a free case review.

Many of the problems with the original pacemakers have been solved through the development of better batteries, titanium shells, semiconductor technology and more durable and flexible pacing wires. The solutions to these problems were innovative, but it was advances in electronic circuitry, which also spurred concurrent dramatic advances in the electronics field, that revolutionized electronic pacing.

The first pacemakers were essentially electronic metronomes. They just tick-tocked until they stopped, like the grandfather's clock in the old ditty, never to go again. With the advent of integrated electronics, however, pacemakers suddenly got smarter: They became *programmable*. While the first pacemakers sent out a pulse with exact, dumb regularity like a lighthouse, the development of integrated electronics made it possible to intelligently combine sensing with pulsing, and the pacemaker became sentient.

The heart has four chambers: two atria (Latin for doorways) and two ventricles (Latin for belly). The atria contract simultaneously to push blood into the right and left ventricles during the first part of the heartbeat (the "lub"), thereby priming the pumps. During the second part of the beat (the "dub"), the much stronger ventricles squeeze blood out of the heart and into the lungs and body. This rhythmical music of the heart is coordinated by the appropriately termed cardiac conducting system, which is what is diseased in many people who need a pacemaker. The modern, smart pacemaker's job is to listen to the heart as it beats, and to step in and pace when one of the chambers doesn't beat in the right sequence. The sensing pacemaker acts like one of those old married couples, where one partner listens vigilantly for the slightest hesitation in the other's recitation and jumps in promptly to fill in gaps in an oft-told tale.

As with music, the normal heart has both slower andante and faster allegro interludes, the latter typically resulting from stress or exercise. The heart beats faster to supply more oxygen to exercising muscles, so rate-responsive pacemakers are designed to artificially increase pacing speed when they perceive the need for increased blood flow, thereby replicating the action of a normal heart.

A pacemaker can sense body movement and breathing, indirect indicators that more blood flow is required. For example, some of the first rate-responsive pacemakers sensed vibration, and the more the pacemaker was bounced around, the higher the rate at which it would pace. Unfortunately, the early versions tended to behave awkwardly, inappropriately increasing heart rate when, for example, the paced patient rested in a vibrating massage chair, operated a jackhammer or bicycled on cobblestone streets—and if you haven't done the last of these, I recommend it. Newer pacemakers have smarter algorithms that are better able to discriminate between artifact and activity. The newest pacemakers are also programmable. A cardiologist can apply a computer to the skin overlaying a pacemaker, interrogate it and reprogram if necessary.

Most recently, pacemaking has been combined with cardioversion and defibrillation capabilities in devices known as Automatic Implantable Cardioverter-Defibrillators or AICDs. AICDs can pace intelligently, deliver an electrical shock to the heart if it goes into a rhythm like ventricular fibrillation, or both defibrillate and then pace if needed. These very smart machines combine both a *sensing* and an *effector* function with a fairly sophisticated electronic brain. Moreover, these little packages represent a radical break from the way medicine has been practiced in the past and a model for the way many medical interventions will work in the future.

The AICD is an *emancipated* medical machine enabled to act *autonomously* to diagnose and then deliver the best possible therapy picked from its built-in library of potential treatments. In traditional medical therapeutics there is always a human provider, typically a physician or a nurse, who analyzes patient data before prescribing a treatment. The AICD, however, senses a problem, analyzes and acts all on its own. Some human programmed the world of possible interventions into the device, but the moment the patient walks out of the electrophysiology lab, the AICD is on its own. If it works, it saves the guy's life. If it doesn't, the guy may be dead. But the data clearly show that AICDs save a lot of lives.

Roger Watson knew the chips were stacked against him from a pretty young age. Both his father and grandfather had died in their forties from a heart attack, so when his wife became pregnant for the first time, he started to think about things like his health, life insurance and a will—things he'd tried to avoid thinking about in the past. Watson's father had been on a sales trip and died alone during the night in a hotel room. He was found the next morning by the hotel's housekeeping staff. He hadn't been ill; he just had a massive heart attack and died. Watson was eighteen at the time, off at college, and the man who had been his role model and with whom he hadn't talked in a couple of weeks just suddenly stopped being there.

Watson decided he didn't want that to happen to his wife or their child, and determined to do something about the things he could control while there was time. He saw his internist and had a thoroughgoing physical and tests, and the results weren't great. He smoked. His blood pressure was higher than it should be. He wasn't obese but he wasn't trim either. He didn't exercise. And his cholesterol was pretty high. He had almost every risk factor for what his doctor called coronary artery disease as well as every other artery disease, and he was 29. They both agreed he needed to turn it around pretty quickly or he'd end up like his father.

It was surprisingly easy for him to quit smoking, now that he had what he thought of as a noose around his neck. He and his wife spent some talking about how to radically change what they ate. Their mantra became "cheese is the enemy." They both got very smart about reading the labels on boxes and cartons and were astonished to find out how many foods contained high fructose corn syrup. Who would have thought, for example, that a sushi salmon roll or a salad topping would contain sugar water? Within a month of his doctor's visit, healthier friends who used to drop in to say hello and to eat came by less often because everything in the Watson house was now low sugar, low salt and low fat—and consequently not what they were accustomed to.

Roger lost his gut, got his blood pressure into a range that satisfied his doctor with a little medicine, and his bad cholesterol came down a hundred points. And then the baby was born. Neither of the Watsons was prepared for how much of a lifestyle change the baby imposed on them, but they muddled through with the help of relatives and friends, and six months later everything was going fine. The baby was healthy, and Roger had started to jog a little, which felt good both as a stress-reliever and a way to keep the weight off.

One Tuesday evening around 5 PM, he headed out in his sweats, saying he'd be back in half an hour, but he hadn't returned by 6, which was when they usually fed the baby—something Roger liked to do himself when he could. By the time an additional hour had passed without word, his wife was frantic; and she called a friend to ask him to check Roger's jogging route, because she didn't want to leave herself lest he call or come home. Twenty minutes later the phone rang. It was the hospital.

Roger had evidently been running alongside the park near their house, where a softball game was underway, when he appeared to stumble and then fall flat on his face. Someone in the stands next to the field happened to notice and screamed because he started having convulsions. Fortunately one of the players was a paramedic who ran over and started CPR while someone else called for an ambulance. The medical team arrived quickly from the nearby firehouse. They quickly determined that his heart was in ventricular fibrillation and shocked him.

It's never really clear how things are going to play out in those first seconds after someone collapses out there in the world. A lot depends on why it happened, although something that drops someone in their tracks, as in Roger Watson's case, is almost invariably the heart. Then it's a matter of whether or not they get CPR; if so, how quickly and effectively; and how long it took for the ambulance to get there and other intangibles. And these are the moments medics live for, when they're on the front lines of a medical crisis and their actions will make a big difference in whether someone lives or dies. Everything happened just right for Roger that evening, and the first time they defibrillated him, his heart went right back into a normal rhythm, and he started to wake up immediately once the blood was flowing to his brain.

The paramedics loaded him up in the van and sped off to the hospital, where it quickly became apparent that Roger had had what is called a silent myocardial infarction, a heart attack without any premonitory symptoms. Basically, while he was running, blood stopped flowing through one of the three arteries to his heart, which caused immediate heart muscle damage followed by fibrillation. The CPR provided enough blood flow to keep his brain from being damaged until they shocked him; but he hadn't called home immediately because as soon as he arrived in the emergency ward, he had another fibrillation episode, followed by two more.

The doctors were able to defibrillate him each time with one electrical shock, but Roger was dangerously close to death until they gave him an intravenous drug that stopped the episodes. Over the next couple of weeks, it became clear that, despite Roger's best efforts, he had gone through a big heart attack and that some major medical interventions were in order. One of the interventions the doctors recommended was the AICD. They said that it was better than long-term drugs as a way to treat a now-clear and persistent heart rhythm problem. The Watsons agreed.

Like many people who suffer large injuries to their heart, Roger Watson went on to develop progressive heart failure over a couple of years and became progressively more debilitated. He had to sit on the sidelines as his child grew up, because he wasn't able to lift things or keep up very well. He couldn't drive because his AICD shocked him periodically at unexpected moments, but he realized that each shock was another lease on life, one that hadn't been available to previous generations. Sometimes he felt faint and passed out, and his wife had to tell him what had happened; on other occasions he felt what he described as a "kick in my chest." Eventually, the doctors put him on a heart transplant list, and one afternoon he received a call from the transplant coordinator saying a heart had become available.

His wife drove him to the transplant center, and a couple of hours later, he lay on an operating room table, ice cold, with his chest wide open, a cavity where his heart had been, and the two silvery metal wires of the AICD that had kept him alive for the past several years peeked out of a cut blood vessel. The surgeons sewed in his new heart, and the last thing they did before closing his chest was to take out the metal case containing the brains of the old AICD because he no longer need it. The next morning Roger Watson awoke to see his wife and child at his bedside. He realized at that point that he would live longer than his own father and grandfather, and was grateful.

AICDs have been around for a while. They've saved a lot of people's lives and have become substantially more sophisticated, even as they've been shrunk to a tenth of their original cigarette-pack size. And like pacemakers, which originally required a big operation, AICDs are now implanted under local anesthesia. The big breakthrough, however, is one that isn't discussed much: AICDs are little doctor-bots. They make life-and-death medical decisions all by themselves, albeit after having been programmed by very smart physicians. This kind of small but smart autonomous medical device will have many applications in many other diseases in the future, such as the management of blood suger control.

In the traditional treatment of diabetes, a glucose test is performed and then the doctor *thinks* . . . before he prescribes. Newer therapies are increasingly automatic, albeit to a degree. Take the insulin pump, for example. This miniature, beeper-sized pump dispenses insulin in both a continuous and bolus mode, allowing it to mimic the treatment function of the human pancreas. In today's devices the patient must still enter a blood glucose level into the pump manually and must therefore periodically measure blood sugar. However, *continuous* blood glucose monitors have recently become available that measure blood glucose levels automatically every few minutes, much more frequently than a patient could feasibly do manually. In the very near future, these monitors will be married directly to the pump, and both will be connected to a semiconductor brain that will automatically control delivery of the drug—another example of a doctor-bot.

It's not hard to imagine automatic dispensation of a number of other treatments. Our blood pressure, for example, goes up and down depending on such things as how stressed we are, whether we're awake or asleep, and how much salt we've eaten in the last 24 hours. We only have our blood pressure measured when we go to the doctor's office, and some people's pressure goes up just at the thought of the trip, a phenomenon known as white coat hypertension. Blood pressure medications are taken once or twice a day, and once prescribed, aren't usually adjusted for months. So while blood pressure varies from minute to minute, we treat high blood pressure with drugs that work over hours and we measure it monthly at best. It would be much better to measure pressure from minute to minute and treat it quickly when warranted. What if we had a device that sat somewhere in the body continuously sensing pressure and communicating with a pump that would then deliver exactly the right amount of drug into the bloodstream to keep the pressure precisely where it ought to be at all times?

Pacemakers and AICDs measure and treat electrical problems in the heart, but what if we had a pacemaker that could be implanted in the *brain* where it could sense an impending seizure and discharge an electrical signal to abort it? This may sound familiar to some readers because it is the premise of Michael Crichton's 1972 science fiction book, *The Terminal Man*. It might even be a promising reality for others because this exact technology is currently being studied in real patients who experience hard-to-control seizures.

The potential promise of various kinds of brain stimulation is enormous. Miniature brain stimulation devices or pacemakers that deliver targeted electrical pulses to selected nerves and deep centers in the brain have been successfully used to treat seizures, tremor, Parkinsonism and pain. A variety of other brain-based syndromes are potential targets for neural stimulation therapy, including depression and obsessive-compulsive disorder.

In fact, deep brain stimulation was recently used to awaken a middle-aged man from the coma he'd been in for six years. Over a period of time following the initiation of treatment, the patient progressed from a minimally conscious state, in which he was only able to intermittently move his eyes or fingers, to the point that he could talk, eat and play cards. The physicians who designed this therapy implanted miniature electrodes very precisely in a small but critical brain center and stimulated them with a pacemaker.

Medical environments are increasingly packed with microprocessors, which while they have "little minds" are consistent and reliable. To give you an idea of the role microprocessors play in the care of patients in a modern environment, let's take a virtual walk around the operating room where our team performs cardiac surgery every day.

As you walk into the operating theater, the most obvious examples of microprocessors at work are the computers used by the nurse and anesthesiologist. The nurse uses a personal computer to track and order supplies, with software that bills the insurer and triggers inventory management, just like the cash register transaction at a modern store. Another operating room control program acts like flight control software, informing receiving areas about when this patient will arrive and when the next patient needs to be ready. The anesthesiologist keeps the anesthesia record of vital signs, blood loss, fluids and administered drugs on *his* computer, and the anesthesia machine is run by microprocessors. All of this was handled manually using pencils, phones, and mechanical devices just a few short years ago; things are far more efficient today.

The patient's vital signs—the heart rate, blood pressure and oxygen levels—are measured by a series of patient monitors stacked on top of the anesthesia machine, all of which are full of microprocessors. The infusion pumps that administer intravenous medications are microprocessor-controlled and have a built-in library of all the drugs and their safe infusion rates, as well as a medication calculator to determine the correct dose rate for a patient of a given height and weight. Most of the drugs an anesthesiologist might need to give during the course of a case are kept in another computerized, password-protected cart that keeps track of what drugs were given to which patient and by whom. When the machine dispenses a given drug, it automatically signals the pharmacist to restock. The pumps of yesterday had no safety stops, the drugs were made up by hand by busy, distracted anesthesiologists, and we tended to run out of things not too long ago.

We use an ultrasound machine to watch the heart as it beats during heart surgery, and the machine is built on a personal computer. The heart-lung bypass machine that controls blood flow when the heart is stopped has, not surprisingly, a combination of manually and digitally controlled functions. Sometimes we monitor the brain and spinal cord with another set of computers. Now, we can detect problems as they occur and do something about them, rather than after the patient woke up, or not, when it was too late.

We used to have a pile of beepers on a table somewhere in every operating room, so that the circulating nurse could handle messages while the surgeons that belong to the beepers were scrubbed. Nowadays, the beepers have been replaced by a pile of cell and smart phones, each with their distinctive little personalities. Frequently during the course of an operation, one of these phones will sound off with whatever ring-tone its owner selected. One friend of mine selected the opening stanza of "It's Been a Hard Day's Night" for his phone and he gets a lot of calls. It's a great little ditty for the middle of a long operation at three A.M.

As played by George Harrison, the distinctive, sustained opening chord of the Beatles' classic song has been variously identified and the subject of a lot of speculation, but I now think of my friend when I hear that chord, and the ring of his phone adds an extra piquancy to procedures in his operating rooms. There's a little mind in every one of those chirpy little smart phones capable of playing any ring-tone you might want, handling phone traffic, showing pictures, searching the Internet, playing games and in some cases even playing movies. And those same little minds may one day be paying attention to our health as we pursue our daily tasks.

All told, there may be as many as fifty or more computer chips in my operating room on a given day. Some are of relatively limited intelligence and only do one job mindlessly, over and over, like the radio frequency identification, or RFID, tags: "Here I am . . . Here I am . . . Here I am." Others are smarter, more communicative and network with like-minded chips using wires or the airwaves.

To date, however, there are still very few emancipated, autonomous computers like the AICD in the operating room, or, for that matter, anywhere else in the practice of medicine. Yet there is no compelling reason that a variety of tasks couldn't be delegated to autonomous doctor-bots just as the critical job of pacemaking and defibrillation has been to pacemakers. For example, it is completely feasible to design a sort of artificial anesthesiologist (like the insulin pump that functions as an artificial pancreas, as described earlier in the chapter), by linking a smart infusion pump with a device that measures the depth of anesthesia and programming a microprocessor to "keep the patient asleep." While a real anesthesiologist's job is much more complicated than this, feedback-controlled processes like this one will be increasingly integrated into health care in an ancillary role in the very near future as software and hardware evolve to assist nurses and doctors.

While the prospect of automating medical treatment may seem foolhardy to some, consider the modern autopilot used on airplanes for takeoff, approach and landing, as well as for routine flight. Autopilots use redundant sets of microprocessors that *sense* the aircraft's position using inputs such as roll, pitch, yaw, altitude, latitude and longitude, and *effect* changes to the aircraft's engines and flaps. Modern autopilots are so reliable and so good at this that they can land an aircraft safely and smoothly in what amounts to zero-visibility weather, the kind of weather John Kennedy ran into on the way to Martha's Vineyard on the night he died. Keep in mind that planes can carry hundreds of people and that *human* error is the most common cause of aviation accidents. Doctors perform surgery on only one patient at a time.

Moore's law of computing suggests that we can expect the processing power, storage capacity and intelligence of electronics around us to increase more or less inexorably for the near future. Miniaturization is driving this trend: As circuits and memory get smaller, we can cram more stuff into tinier machines. Arne Larrson's first pacemaker had two transistors and his last had 500,000 and was a lot smaller.

Within the next ten years, miniature medical devices for personal use will become widely available. Some may take the form of medical jewelry that will measure things such as heart rate, blood pressure, hydration level or blood oxygen levels, or perhaps you'll wear a medical undergarment that measures stress through levels of antioxidants in your sweat.

Miniature subcutaneous medical devices, like today's pacemakers, will soon become sufficiently robust and functional that a healthy individual may well choose to undergo the inconvenience of the implantation procedure. The implanted chip might carry medical records, monitor vital signs or do both. A few hundred people have already had an RFID chip placed under their skins with information linking the recipient to a central medical record repository. Hewlett Packard has recently developed a smart treatment patch with basic electronics and a power source that can be programmed to give one or more medications in a timed fashion through microneedles in the skin, totally painlessly. The patch is only an inch square and 3 millimeters thick. One day soon, an implanted medical microchip may determine that its bearer's blood sugar is alarmingly low and call him on his cell phone to do something about it.

The modern explosion in available processing power has helped us develop the electronic microscopes and computers that investigate small things such as viruses, prions and the human genome with ever greater precision. They are the modern equivalents of Leeuwenhoek's lenses. Microprocessor-based instruments effectively allow us to become Alice in Wonderland by scaling our senses down to micro- and nanoscopic levels.

A demonstration project by Japanese engineers dramatically demonstrates our ever-growing mastery of little things. Working with a team of artists and physicists at Osaka University in 2001, Dr. Satoshi Kawata etched in plastic a beautiful sculpture of a rampant bull, using twin intersecting laser beams as chisels. The bull is the size of a red blood cell, but it is rendered in exquisite detail—arrested in motion, horned head to one side, tail a-twitch. Unfortunately for most of us, it can be seen only with a scanning electron microscope, like prions. Similarly, Cornell University scientists created a nanoscale guitar in the late 1990s with tiny strings. The strings could actually be plucked, like those on George Harrison's guitar, but the resulting chord would, sadly, fall outside the range of human hearing.

While Kawata's nano-bull is a self-sufficient work of art, the laser-based sculpting technology used to create it is primarily being researched as a way to create nanoscale medical machines. One of the scientists from Osaka said in 2001 in the journal *Nature*: "We dream that this bull pulls a drug cart through the blood vessels." Many of the most dramatic medical advances in the next decades will occur at the cellular, molecular and even atomic levels.

In a 1657 letter to fellow scientist Robert Hooke, Isaac Newton wrote, "If I have seen further, it is by standing upon the shoulders of giants." Newton was referring to the debt he owed to his astronomical predecessors Kepler and Galileo. Stephen Hawking stands in turn on Newton's shoulders. Just as our human predilection to be "curiouser and curiouser" has drawn us out ever farther into the cosmos, it has also compelled scientists like Leeuwenhoek, Prusiner and Kawata to delve ever deeper into the stuff of which life is made.

Hawking refers to an elderly woman's statement about turtles in the opening sentence of *A Brief History of Time;* and it turns out that the story behind the turtles, which has various versions, perversely parodies Newton's statement about his forbearers, and by extension his followers. At the conclusion of a lecture by a famous scientist about a theory—some say cosmology, others evolution—an elderly lady at the back of the audience says she *knows* that the world is actually a flat disk supported on the back of a tortoise. When the scientist responds with a smug smile, "But what is the *tortoise* standing on?" the lady is said to have answered, "You're very clever young man, very clever, but it's turtles all the way down." Richard Feynman's 1959 essay, "There's Plenty of Room at the Bottom," is generally credited with launching the field of nanotechnology, which I'll talk about in the next chapter, and implicitly describes what might be under the turtles.

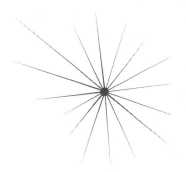

CHAPTER 9

AT THE BOTTOM

When the *Challenger* space shuttle blew up in 1986, I was standing in a patient ward at the Stanford University Hospital that had four separate televisions tuned to that morning's launch. It was a historic flight because one of the crew members was Christa McAuliffe, a secondary school teacher, and across the nation American school children were watching. Barely a minute into the flight, the shuttle's contrail suddenly ballooned and then split in two, and the *Challenger's* velocity slowed abruptly. The television announcer was understandably at a loss for words at first and then managed something about a "major malfunction," clearly as uncertain as the rest of us about what we were seeing. But everyone realized that something catastrophic had happened.

Two weeks later, during presidential commission hearings, Nobel Prize–winning physicist Dr. Richard Feynman demonstrated the exact reason for the shuttle's disintegration. Feynman, who led the inquiry, dropped a piece of the rubbery substance used to make the shuttle's booster O-ring seals into a glass of ice water after screwing on a C-clamp to pinch the material in the middle. The O-rings were designed to act like gaskets between adjacent fuel tanks and to prevent fuel leakage. Despite NASA engineers' claims that the rings were designed to be elastic and flexible even at low temperatures, Feynman, who has been called the "Great Explainer," took a chilled O-ring sample from the water, removed the clamp and showed that the ring material remained deformed and inelastic. The molecules in the seals had changed their behavior in the cold.

Pictures of *Challenger* on the morning of the launch actually show icicles on the gantries and fins, and when the film footage of the liftoff was analyzed frame by frame, it was clear what had happened. At the moment of ignition, the pressure in the booster fuel tanks increased dramatically, and a little puff of gas can be seen escaping at one of the joints where an O-ring sits. It took almost a minute before the full consequences of the O-ring failure became evident; it was the incomplete seal between two booster rockets due to the loss of flexibility in the rings at low temperatures that caused the death of the *Challenger*'s passengers. Sadly, it became apparent during Feynman's investigation that the potential vulnerability of the O-rings at low temperatures had been under discussion the night before the *Challenger* launch.

Feynman won the Nobel Prize in part for his discoveries about subatomic particles, but he was also a gifted photographer, clever prankster and scientific visionary. He worked on the Manhattan Project early in his career, but eventually settled at Caltech, where he gave a now famous lecture on the possibilities of "manipulating and controlling things on a small scale." The lecture was given to members of the American Physical Society, and entitled "There's Plenty of Room at the Bottom"; it laid the theoretical underpinnings for an entirely new scientific field, in which he proposed that molecular-scale, man-made machines and materials would revolutionize the future of industries as diverse as computer science and medicine. The subtitle of the talk was: "An invitation to enter a new field of physics," and that field has come to be what we now know as nanotechnology.

There was no elderly lady heckler like the one in *A Brief History of Time* on the day Feynman presented his remarkable thesis, but the members of the audience could have been forgiven for thinking: "But it *is* turtles all the way down." Feynman was talking about manipulating molecules and atoms in a way that must have seemed inconceivable to many. Nevertheless, Richard Feynman knew that there was something important under the turtles. His talk discussed the possibility of manufactured nanoscale computers, memory storage, self-assembling machines and biologically active devices. As he said in reference to medicine, "Biology is not simply writing information; it is doing something about it. Many . . . cells are very tiny, but they are very active; they manufacture various substances, they walk around; they wiggle; and they do all sorts of marvelous things—all on a very small scale. Also they store information. Consider the possibility that we too can make a thing very small which does what we want—that we can manufacture an object that maneuvers at that level!"

Feynman was prescient, because nanotechnology is in the process of revolutionizing the way we manufacture and use materials in a variety of industries. Nanoscience has already led to the development of novel manufactured substances such as metals, polymers and composites that make traditional materials, such as those used for the *Challenger* O-rings, completely obsolete. For example, the properties of many elements change dramatically when they are manufactured as nanoscale particles. Materials that are ordinarily opaque, such as copper, become transparent; solids such as gold become liquids; insulators such as silicon become conductors, critical to the semiconductor industry; poorly conductive materials become superconductive; inert metals such as platinum and gold become catalysts; and materials that don't ordinarily burn, such as aluminum, become combustible.

As we'll see, these novel properties of nanoscale materials are in the midst of changing the diagnosis and treatment of infections, cancer and autoimmune diseases. Bioengineers are in the process of developing very creative approaches to the diagnosis and treatment of these diseases using little biomechanical machines that are partly biological and partly manmade. The remarkable ability of these machines results from their ability to interact with cells at the molecular level and alter the ways in which cells identify and communicate with one another. In other words, the biomechanical machines operate on the same scale as the molecular machinery of the cell.

Many of the patients I took care of at Stanford in the late 1980s had autoimmune diseases in which the body's watchdog cells turn on their owner. There are a whole host of these syndromes in which the body's immune cells lose the ability to recognize and distinguish between the self and non-self. At least one of the four patients in the ward where I watched *Challenger* explode that morning was a lupus patient, as Stanford was a referral center for the diagnosis and treatment of patients with this complicated disease, sometimes called "the great imitator" for its propensity to mimic other diseases.

More common in women than men, lupus erythematosus derives its name from the characteristic facial rash found in many patients. Lupus is the Latin word for wolf, and patients with the disease usually have a butterfly-shaped red rash across the nose and cheeks. Said to resemble a wolf's bite, the rash often looks like a bad case of acne. Many of the more systemic manifestations of the disease, such as arthritis, can be mistaken for some other medical condition, and it may take several years of waxing and waning symptoms as well as many frustrating medical visits before the correct diagnosis is made. While we don't completely understand the disease, the fundamental problem seems to result from nanoscale cellular debris.

Marianne Bowden was a generally healthy teenager who developed what seemed like a typical viral syndrome about three years before she was ultimately admitted to the hospital, critically ill, as a diagnostic mystery. One winter she had a high fever, bad headache and sweating episodes that lasted for several days, all of which was attributed to something "going around." However, unlike other kids who had gotten the same thing around that time, she didn't ever really get better. Her appetite diminished, and as the following spring progressed to summer, she never got her energy back. She also noticed that her hair was thinner, and it seemed like there were wads of it in her hairbrush. At one point she developed an ugly rash on the back of her legs. Her doctor attributed her low red blood cell count to what he called a poor diet (although she felt she ate perfectly adequately) compounded by menstruation, and prescribed iron pills.

Over the next two years, Bowden had a variety of fleeting, nonspecific symptoms including recurrent migraines and muscle aches—she never felt quite well but was never sick enough to trigger real concern on the part of her doctor. Different joints ached at various times, and she was completely unable to move one morning a day after a family trip to a beach house, because her neck and back were so painful and stiff. She lay on a couch for a couple of days before she felt well enough to get up. At one point she developed a bad sore throat with painful mouth ulcers, which was again chalked up to a viral process by her doctor. Her fingers suddenly started to turn white in cold weather or when she handled cold objects like ice cream. This is a fairly common, nonspecific symptom called Raynaud syndrome that can occur in the absence of systemic disease, and it was labeled as such by her doctor who didn't pursue it further.

Bowden was depressed about the fact that she felt chronically unwell, and her frequent visits to the doctor in the absence of some demonstrable, treatable medical problem left both of them with a sense of mutual dissatisfaction. He felt she was hypochondriacal and occasionally hysterical, and included chronic fatigue syndrome and fibromyalgia in the list of her diagnoses. She felt unattractive, her face was always breaking out, her hair fell out in clumps and she carried herself like an old person—she *knew* something was wrong that he hadn't yet put a name to.

The night before she was finally hospitalized at Stanford, Bowden developed a high fever and chills and what she thought was a migraine. Her thought processes became more confused as the evening wore on and a roommate eventually took her to the emergency room. Her initial laboratory tests were alarming. She was clearly disoriented and unsure about the date or place or even what was going on around her. She appeared to be hallucinating, hearing and seeing things that weren't there. Her lips and tongue were swollen, and she was coughing up bloody spit. She had a bright red rash across her cheeks. A chest x-ray showed fluid around her lungs; blood tests showed that her red and white blood cell counts were low and that her kidney function was abnormal. While she was in the emergency room awaiting admission to a hospital bed, she had a full, grand mal seizure.

By this point, it was clear to everyone, including her chagrined family physician, that she was very sick, with a very real and multiorgan illness, and it didn't take long for her doctors to identify her disease as a bad flare-up of systemic lupus erythematosus. She was admitted to an intensive care unit and started on a treatment regimen designed to shut down what could fairly be characterized as a civil war or riot going on inside Marianne's body, where her white cells were metaphorically revolting against their neighbors. In retrospect, of course, many of her symptoms dating back to that initial episode three years earlier were entirely typical for lupus, but her doctor can be forgiven for having missed the big picture because lupus signs and symptoms are so nonspecific.

Certain diseases, like certain people, call attention to themselves and are easy to diagnose. For example, if a patient turns yellow and feels tired, you can be pretty sure that there is something wrong with the liver. Lupus, on the other hand, can run under the radar for a long time before it is diagnosed: If a patient is chronically tired and nothing shows up in routine tests, lupus is not usually the first thing that comes to a doctor's mind.

Put simply, lupus is a disease wherein two types of white blood cells, the B- and T-lymphocytes, fail to make that very important distinction between self and nonself. T-lymphocytes may attack normal tissue directly or secrete chemicals causing inflammation and damage to normal cells; B-lymphocytes secrete antibodies that attach to normal tissue leading to its destruction or the formation of antibody–antigen complexes. Under normal circumstances, antigens are found on bacteria and viruses, and the antibodies act to flag the foreign tissue for destruction. In lupus, however, the antibodies and T-cells incorrectly identify normal DNA and RNA as foreign and the disease becomes an ongoing cellular insurgency, with some degree of constant low-grade inflammation punctuated by flare-ups like the one that finally brought Bowden to a diagnosis. One of the diagnostic tests for lupus uses a fluorescent dye to label antibodies, and in lupus-positive patients you can see the labeled, glowing, abnormal antibodies pasted up against the windows of the nucleus of normal cells, trying like zombies to get at the DNA inside. One theory to explain lupus holds that abnormal lymphocytes may come to react against the self after exposure to DNA and RNA debris from cells that died a normal death but were not cleaned up expeditiously. In this theory, the B- and T-lymphocytes suddenly go haywire, becoming overzealous sanitation workers who break into homes and carry off the functional things lying around the house in the mistaken belief that they are just as disposable as their broken equivalents at the curb.

The symptoms of lupus are quite varied and the severity can wax and wane, although there are clear triggers for flare-ups. Mouth, nose and vaginal surfaces can develop ulcers and many patients experience hair loss. These are all manifestations of the autoimmune reaction against various skin surfaces on and in the body. Joint and muscle pain are very typical, and while patients may not have joint destruction, as they do with rheumatoid arthritis, joints in the hands and feet can become swollen and stiff. The blood of lupus patients can be abnormal in many ways. Anemia and low clotting and white blood cell counts are common. Despite the low clotting cell counts, patients are actually *more* prone to the development of blood clots in the legs or lungs. Lupus can also cause inflammation in the heart, lungs, kidneys, liver and, as with Bowden, in the brain.

While we have a very good understanding of lupus—we know something about the genetic abnormalities associated with it and we know many of the things that trigger flare-ups—unlike many other diseases, we don't really have a clear understanding of the roots of the disease, and it is quite likely that more than one cause may trigger a common sequence of results that we lump under the heading of lupus. There are clear genetic aspects to lupus, as the disease tends to run in families. There are also a variety of environmental precipitators, including drugs, such as antibiotics and antidepressants, hormones, infections and sunlight. For example, birth control pills or hormone replacement therapy have been shown to trigger outbreaks of the disease. The virus that causes mononucleosis, called the Epstein-Barr virus, can activate B-lymphocytes and thereby trigger the disease. Ultraviolet light, which is an element of sunlight, can cause the development of a lupus rash, and it is not uncommon to hear patients describe the development of a flare-up of symptoms, as Bowden did, after a trip to the beach. Ultraviolet light does something, although it isn't clear what, to activate immune cells against normal skin cells, causing the development of a rash on sun-exposed areas. That inflammatory reaction may spill over into the rest of the body in some patients who have generalized, multiorgan flare-ups after brief exposures—even as short as half an hour—to the sun.

A somewhat surprising aspect to the disease is the fact that flare-ups can be triggered by low *antioxidant* levels. An easy way to think about oxidants and antioxidants in the body is to think of rust. Oxidation can cause damage to cells in the same way that it causes metal to rust; and antioxidants like vitamins C and E act like rust-inhibitors to prevent cell damage. Antioxidant-rich diets or diets with selected nutrients can keep lupus in check by keeping someone's cells in good working order. Good nutrition and consequent high antioxidant levels appear to act as an anti-inflammatory, which is probably a good lesson for all, even the healthy among us. Similarly, exercise seems to ameliorate the fatigue and improve the well-being of lupus patients.

Most of the standard treatments for lupus are anti-inflammatory agents, such as aspirin, acetaminophen and steroids and immunosuppressive drugs.

These drugs don't treat the specific causes of the disease; they just act more generally to suppress inflammation. If you think of the disease as a sort of cellular insurgency, then anti-inflammatory agents and immunosuppressives are like a curfew, or random military patrols designed to keep the disease off the streets. Bowden, for example, was treated with high-dose steroids shortly after her arrival in the intensive care unit; the symptoms, including her neurological abnormalities, lung inflammation and kidney abnormalities resolved after a few days. High-dose steroids are like a troop surge, and they typically work—inflammation subsides and the patient feels better. The problem, of course, is the toll on the patient.

Long-term, high-dose consumption of steroids causes obesity, brittle bones and, most importantly, a weakened immune system; rheumatologists—the doctors who specialize in the treatment of autoimmune diseases—do their best to prescribe patients the lowest possible doses of steroids or get off them altogether. Unfortunately, the other immunosuppressives used to treat the disease increase susceptibility to infections as well. Although patients are clearly much better off today than fifty years ago, lupus is still debilitating, necessitating significant changes in lifestyle as well as continual medical vigilance. It would be nice to have a cure.

There are a number of experimental treatments under evaluation that go beyond the blunt force approach of steroids and immunosuppressives. For example, bone marrow transplantation with a patient's own stem cells could be a solution. This approach has been used for various leukemias, multiple myeloma and breast cancer, and basically involves taking a sample of the patient's primitive bone marrow cells and growing them in a laboratory. The patient is simultaneously treated with high doses of chemotherapy or radiation to eradicate the existing white blood cells in the body in a nondiscriminatory fashion, including both the good and the bad actors. The stem cells are then used to regenerate the barren marrow, essentially repopulating it. The major problem with bone marrow transplantation is the extremely perilous interval during which the body is without white cells. A much more appealing approach would be to find some way to handle the battle between unruly white blood cells and normal cells at the street level, so to speak, with something analogous to community policing or the zero-tolerance approach many cities have recently adopted to minor crimes like graffiti, jaywalking and public urination. The idea would be to block the effects of lupus at the cellular or molecular level.

Nanomedicine is a blanket term that covers a wide variety of potential treatments, some of which have already been deployed, and many more of which are purely fantasy at this point. When Feynman gave his nanotechnology lecture in 1959, he indulged in a thought experiment to describe how one might go about engraving the entire *Encyclopedia Britannica* on the head of a pin, so let's indulge in the same process by imagining a nanomechanical cure for lupus.

One of the theories about the cause for lupus suggests that patients have a problem with a process known as *apoptosis*, the normal, programmed death of cells in the body. Apoptosis is ordinarily a way for the body to keep house and eliminate "rusted" cells in which the DNA is damaged, possibly by exposure to ultraviolet light or by oxidants and free radicals. These damaged cells ordinarily commit suicide, honorably and in a tidy fashion; under normal circumstances, janitor white cells, called phagocytes, then swoop in and clear up the dead cells.

The apoptosis lupus theory holds that something goes wrong with the apoptotic process or that the cleanup isn't handled properly and pieces of the dead cells' DNA and RNA are left lying about. B- and T-lymphocytes, which constitute the body's cleanup crew and don't like messes at all, then become confused and frantic and start to develop an immune response to remove these *fragments* of normal cells. The agitated lymphocytes then, inappropriately, start to chew on healthy *cells* just because they happen to contain DNA and RNA that looks like the fragmented stuff they just finished clearing away.

Presuming the lupus patient's cells continue to die gracelessly, this theory provides a good mechanistic model for the chronic low-grade autoimmune, or anti-self, lymphocyte activity in lupus. You can explain the intermittent symptom flares by invoking such things as intermittent periods of enhanced cell death due to sun exposure or excess free-radical intake, from, say, a quick cigarette or a fat-laden pastry. Proceeding, then, with our thought experiment, an ideal way for a nanomedicine technologist to set about curing lupus would be to fix the broken apoptotic process so that lymphocytes are no longer exposed to cellular debris and therefore no longer feel a need to beat up on normal cells.

Feynman predicted a day when we might be able to reduce the size of the automobile, the computer and mechanical tool to the size of a molecule. His talk inspired a science fiction movie of the 1960s called *Fantastic Voyage,* in which a submarine and its crew are shrunk to the size of a cell and then injected into the bloodstream of a dying scientist on a mission to remove a blood clot from his brain with a surgical laser. During the perilous mission, antibodies attack the female lead, Racquel Welch, eventually covering every inch of her tight, white wetsuit, and the submarine itself is subjected to the less-than-friendly attentions of a giant white blood cell. The immune system seems set on killing off the submarine and every one of its occupants, all of which are non-self.

While the movie did, in fact, seem totally fantastic at the time, a number of recent advances suggest that it might eventually be feasible to deploy fleets of nanoscale "submarines" of one sort or another into the circulation with specific, preprogrammed instructions to perform some activity—perhaps kill cancer cells, or deliver drugs to particular locations, or even to sweep up cellular debris. If, for example, you could program nanosubs to look for nucleic acid litter and clean it up, you could potentially cure lupus.

A variety of nanomedicine research projects are currently underway at universities, pharmaceutical and medical device companies. Among the potential nanotechnology applications in medicine are nanoscale cancer-killing machines that can circulate through the body looking for cancer cells and then kill them with a dose of toxin; nanoscale coatings for medical devices; nanosensors that can analyze tissue conditions at the cellular level; nanocontrast agents that can light up specific cells in diseased tissue; nanoscaffolding for the assembly of artificial organs; tiny nanoscale tankers that can carry oxygen to tissues with poor circulation; and nanosurgibots designed to perform cellular surgery. Admittedly, there are a huge number of technical obstacles that must be overcome in order to make nanomedical machines a reality, but meanwhile, nanotechnological approaches to the diagnosis and treatment of cancer will be common within the next several years.

In order to understand the way in which nanoscale applications will work in the human body, it is essential to understand something about nanomaterials. Liposomes, for example, were one of the original nanomaterials. Liposomal drugs are basically cell-sized droplets of a drug coated with a layer of normal human cell-wall material. A wide variety of liposome-coated drugs, including chemotherapies and antifungal agents, are already in use today.

Liposomes are only one of the ways nanotechnology can be used to increase the effectiveness of drugs. Intricate, single-layer carbon structures known as fullerenes come in several shapes including spherical buckyballs and cylinders called carbon nanotubes. It is also possible to make ellipsoids and flat sheets called planes of carbon. Fullerenes of any shape can be used as drug carriers, as structural elements in nanoscale machines or attached to strands of DNA or RNA to ensure very specific binding. For example, one way to deliver drugs or diagnostic agents to an individual tissue area, such as a cancer or an inflammation, would be to encase that drug in a fullerene cage and attach a piece of DNA or RNA to the cage. The DNA or RNA would attach to a specific cancer, for example, before the cage releases its payload of chemotherapeutic drug.

Nanoshells are similar to the carbon-based fullerenes but have one material, such as silica, on the inside with a coating of a metal, such as gold, on the outside. Like fullerenes, nanoshells have also been proposed as a cancer treatment approach. The nanoshells can be constructed with a coating of cancer-specific antibodies so that they attach only to cancer cells. Once a cancer cell is covered with nanoshells, the shells can be activated with light or sound waves individually tuned to a wavelength that resonates with the metallic coating, thereby killing the cancer cell and leaving the surrounding cells intact.

Dendrimers are branched nanoscale structures that act like little Velcro balls. A dendrimer can be coated with one or more of a variety of materials, including cell identification tags, enzymes, dyes and drugs. Like fullerenes and nanoshells, dendrimers can also be equipped with DNA, RNA or antibodies so that they can find cells of interest, such as cancer cells, and then label or kill them without stimulating the immune response in the process. Alternatively, a dendrimer might be coated with some molecule that makes it attractive to cancer cells, such as folic acid, and thereby gain entry to the cell. In this approach, the folic acid coating acts like a sting operation. Cancer cells love folate, so they swallow the dendrimer and, once inside, the internal pH of the cell activates an enzyme or releases a toxin attached to the Trojan-horse dendrimer that kills its host. When its task is complete, the dendrimer will be cannibalized into its constituent proteins.

Quantum dots are tiny, nanoscale crystal semiconductors that can emit light in a variety of colors when irradiated with ultraviolet light. The quantum dots can be bound to cancer-specific DNA and used to label malignant cells. Similarly, gold nanorods are another form of semiconductor that can be attached to DNA and used to identify cancer cells or to show blood flow far more precisely than current methods permit.

Superparamagnetic nanoparticles are iron-based molecules that develop magnetic properties in a magnetic field like an MRI. When attached to antibodies, these nanoparticles can be used to fish biological molecules out of a solution like blood by applying a magnetic field, which then pulls the iron, antibody and molecule complex toward the magnet.

Each of the these nanomaterials can be thought of as a building block for medical applications of the present and immediate future. Their common feature is the fact that they are essentially the same size as the key elements of the biological cell, and they can therefore be engineered to interact with cellular components. A nanometer is one-billionth of a meter, and DNA is only 2 nanometers wide. Proteins range in size from 5 to 50 nanometers, and viruses from 75 to 100 nanometers. By way of comparison, a typical lymphocyte or phagocyte is 10,000 nanometers wide.

The fact that nanomaterials are similar in size and scale to cellular molecules makes it possible to design interactions between biological molecules and manmade molecules that wouldn't otherwise be feasible. One way to think of this is to imagine tying two pieces of string together. If one of the pieces of string is a ship's hawser seven inches in diameter, and the other is a length of sewing thread, it's hard to imagine how to tie an effective knot between them in the first place, not to mention the use of such a knot. On the other hand, if both pieces of string are the same size, there are an infinite number of ways to attach the two and a commensurate number of potential applications. We are now able to manufacture nanomaterials and tie them to biological materials of the same scale, opening a world of new medical possibilities.

A counterintuitive aspect of nanoscale mechanics relates to the way that properties such as gravity, friction, conductivity and shear or tearing forces change when acting at the submicroscopic level. In order to get a feel for just how different things are in the nanoscale world, it is worth taking a side trip to discover how those cute little lizards, geckos, walk across a ceiling or scamper up a pane of smoothly polished glass. This seemingly amazing talent remained unexplained for a long time, but by peering under the gecko (albeit long and hard with high-tech equipment), scientists ultimately answered some questions about what lies underneath. To many traditional scientists, suction seemed the most likely explanation, although glue, Velcro-like hooks and static cling were also entertained. Suction won't work in a vacuum, and it turns out that a gecko's foot still does. After a lot of what might superficially seem like frivolous research, scientists from the University of California at Berkeley finally discovered the answer, which happens to have very specific real-world and, unexpectedly, medical applications.

The scientists found that geckos have two million or more hairs, called *setae,* on each toe; the end of each of these setae is further divided into hundreds to thousands of structures called *spatula.* Combined, the setae and spatula create an enormous contact surface area between the gecko toe and the surface to which the lizard is clinging. Interestingly, microsetae are also used by certain insects to trap air bubbles and thereby allow them to walk on water. Geckos, it turns out, walk in such a way as to roll their toes onto the surface and then peel them off with each step. While the toe is in contact with a surface, weak, nanoscale interactions, called van der Waals forces, operate to electrically glue the toe to the glass underfoot. These forces are so effective that geckos can actually hang on to a ceiling with a single toe. Unlike real glue, however, gecko glue is not sticky—it works because of the attraction between negatively and positively charged molecules all along the surface area between the toe and the surface, whether glass or stucco.

The whimsical discovery of gecko glue led to a search for functional, man-made adhesives for a variety of applications, including bandages. A team from MIT recently announced the development of a waterproof dressing that could be applied directly to wet, bleeding tissues in the body, such as intestines, the heart or lungs. These bandages have artificial setae and spatula that are etched onto their sticky surface with a laser and can be impregnated with antibiotics, growth factors or even stem cells to prevent infection and enhance healing. In effect, the dressing can be designed as a nanomolecular scaffolding to stop bleeding and promote healing, just like nature's own.

The van der Waals forces used by geckos are a function of interactions at the molecular and atomic level, and we are now acquiring the ability to harness similar nanoscale forces in a wide variety of medical applications. Friction is another macromolecular property that changes dramatically at the nanoscale level, where a single molecule of oil is relatively huge and no longer effective. Light, magnetism, oxidation and electrical conduction all act differently when they operate at the nanoscale level. As a result, it is possible to develop nanomolecular materials that are more porous, better insulators, more conductive, stronger, lighter, more magnetic and less corrosive.

In an earlier chapter I described the seemingly quixotic scientific investigation of the mystery of the squid's beak, very much like that of the scientists who studied the gecko's foot. Both cases demonstrate that the findings of the curious scientists had direct relevance to modern medicine. Nature provides many such answers, such as the best approach to securing a prosthetic joint to bone. Nature has designed roots to secure trees to the soil; the artificial joints of the future may well use a similar nanoscale approach to do away with the need for glue.

Just as it is possible to glue a gecko's feet to glass by molecular binding, it is equally feasible to create strong bonds between the body and prosthetic devices by increasing the surface area on the implant, allowing bone to grow roots into the implant. Joint implants have traditionally been glued into place with acrylic glue called polymethylmethacrylate, which can become infected or provoke a fibrous reaction between the bone and the prosthetic joint, eventually causing the joint to loosen. The bond between bone and an implant coated with nanoparticles will not require glue and yet be just as strong as the original bone, because the body can grow roots into the nanoparticulate soil. Similarly, nanoceramics are under development that can be used as bone substitutes or even as a kind of epoxy to quickly repair broken bones.

Several types of artificial nanomolecular muscle have been developed that will power the prosthetic limbs of the future. One type of muscle uses platinum-coated wire that expands and contracts depending on the surrounding temperature. Another artificial muscle technology relies on electrical interactions between adjacent carbon nanotubes when they are subjected to an electrical force. Carbon nanotube muscle expands when a charge is applied and contracts when the charge is removed. A third type of muscle relies on the properties of a special class of plastics called conducting polymers that contract in response to an electrical charge. These artificial muscle technologies will be used for medical applications that range from powering an artificial limb to opening and closing tiny valves in implanted drug-delivery devices.

While the currently feasible applications of nanotechnology to medicine are relatively diverse, the range of *potential* technologies is almost limitless, and modern visionaries have proposed medical machines that seem as bizarre today as Feynman's ideas must have seemed to his audience in 1959. Various researchers have described the manmade equivalents of almost every type of cell in the body as well as nanoscale construction of new biological organs from stem cells. It is possible to imagine the development of nanomechanical equivalents of the human cells that sense sight and sound, which could then be used to treat various forms of blindness and deafness. I have already described a theoretical nanoscale garbage scale that could be used to treat or cure lupus. These technologies describe ways that we might apply our growing understanding of the molecular underpinnings of biology to the treatment of a disease. Some nanomedicine futurists have reached *beyond* that mark, however, and proposed nanoscale machines with capabilities that *improve* upon nature.

Robert Freitas has written extensively on the potential applications and challenges of nanotechnology in medicine and has described a theoretical respirocyte, the nanomechanical equivalent of a red blood cell, designed to carry oxygen from the lungs to the tissues and carbon dioxide back to the lungs. As Freitas describes it, in his July 1998 article in *Artificial Cells, Blood Substitutes and Immobilization Biotechnology,* his version would be better than the original and could endow a human with superhuman capabilities. While normal red blood cells are six to eight micrometers in size, the respirocyte is just one micron in diameter; and while the normal red blood cell is basically a specialized, cellular container constructed to carry hemoglobin through the bloodstream, Freitas's respirocyte is a little machine designed to circulate through the blood like a red cell but carry oxygen like a supertanker.

Human red blood cells have what is called a biconcave shape, describing its double dimples, and have the same liposomal bilayer cell wall of all cells. This cell structure is very efficient for the exchange of oxygen and carbon dioxide in the lungs and organs because it maximizes surface area. The biconcave shape also allows the cells to squeeze through capillaries without popping in the process. Freitas's hypothetical nanoscale respirocyte is designed to be smaller than the red cell so it doesn't have to squeeze, and small enough that it can float through the capillaries, which are typically eight micrometers in diameter. The theoretical respirocytes would be constructed of a carbon-based, diamond lattice acting like a tiny scuba tank, which can be pumped full of oxygen or carbon dioxide up to a thousand times the atmospheric pressure. As a result, the respirocyte can carry a whole lot more oxygen or carbon dioxide than an ordinary red blood cell.

The respirocyte would be designed with an onboard nanocomputer and external sensors that could measure the ambient concentrations of oxygen and carbon dioxide as well as an array of what Freitas calls "molecular sorting rotors" that would load and unload gases from the tank to the bloodstream. Like red cells, the respirocytes would load up on oxygen as they pass through the lungs, simultaneously dumping carbon dioxide, then unload the oxygen into the tissues as they pass into the capillaries in the organs and muscles, while pumping the pressure tank up with carbon dioxide released from the functioning cells.

The structural stiffness of the respirocyte would allow it to deliver over two hundred times more oxygen per unit of volume than a normal red blood cell, and Freitas theorizes that the injection of only five cubic centimeters of a saline suspension of respirocytes into the bloodstream could entirely replace the function of all of the red blood cells in a normal adult. While most of us would probably prefer to have normal blood, rather than a bunch of highly pressurized gas canisters circulating through our bloodstream, we'd all probably be a little more interested if we could hold our breath under water or in high altitude conditions for hours rather than minutes. Freitas suggests that a transfusion of a liter of respirocytes would allow you to stay under water, albeit at rest, for four hours, and to sprint at top speed for fifteen minutes without taking a breath. Respirocytes could also have enormous implications in the treatment of disease, where combinations of lung, heart and circulatory problems significantly limit or shorten the lives of many patients. Indeed, a patient's answer to the question, "How many flights of stairs can you manage without getting short of breath?" is a very good way to determine how much oxygen their heart and lungs can produce when needed.

Freitas and others have actually described a vision in which entire fleets of carbon-based, medical nanorobots equipped with computers, sensors and tools would cooperate to cure a disease like cancer or lupus. He suggests that the fleet would then be removed from the body through some natural excretory process, like urination or defecation, or perhaps by exfiltration, like special-forces operatives.

The key to the safety and specificity of any nanoscale treatment is its ability to recognize some unique feature on the target cell and we've already discussed ways in which DNA or antibodies can bind to cancer-specific cell surface identifiers. Bacteria, viruses and tumor cells all have surface features that can act like handles, and the immune system has what amounts to a biometric recognition system that allows it to recognize and destroy the nonself. Regardless of how simple or complex, the effective nanoscale technologies of the future will be selective and individual in their interactions within the body.

The prospect described by futurists such as Freitas is clearly speculative, and the direction and form that nanomedicine will actually take is uncertain. Still, dramatic new discoveries are reported regularly. One of the very first tools discovered by man was the lever, about which Archimedes said: "Give me a place to stand on, and I will move the earth." Nanotechnologists have already designed a nanoscale sensor in which antibodies are attached to a tiny cantilever. When the antibody attaches to a cancer-specific antigen, or a bacterium or virus, the cantilever bends slightly, but sufficiently that it can be detected and therefore act as a diagnostic device. We are now building the nanomolecular equivalent of levers and wheels. Our ever-deeper understanding of the ways that nature has designed her own tools will allow us to make dramatic leaps forward in very short order.

With these possibilities in mind, it is instructive to return to Marianne Bowden whose lupus flare-up was treated with high doses of steroids. She got better but the treatment was palliative rather than curative. Steroids have many side effects and the one that counts in lupus is that they kill lymphocytes. Steroids are used to treat lymphoma for the same reason. If you give a large dose of steroids to someone, the blood count of lymphocytes plummets. On the other hand, HIV also kills lymphocytes, and the most common cause of death for steroid-treated lupus patients and HIV patients alike is the same: infection. Lymphocytes are an essential part of the immune system, and the treatment of lupus with steroids can only be characterized as using a crudely nonspecific, blunt instrument—especially when compared with the typical elegance and precision of the immune system.

Nanotechnology holds promise for energy production, pollution remediation, transportation and medicine; and advances in each field will spur innovation in all of the others. It isn't clear how quickly nanotechnology will revolutionize medicine, but it's a question of when, not if. Consider the changes in the size of computing devices that have occurred in the five decades since Feynman's talk. Significant nanomedical advances, probably in the areas of drug delivery, diagnosis and certain treatments, will almost certainly occur within the next five to ten years. Dramatic changes are on the visible horizon that will almost certainly antiquate many of the current forms of medical diagnosis and treatment.

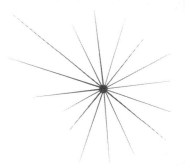

CHAPTER 10

THERE AND BACK AGAIN

January 1 comes about halfway through the winter for people living in the more northerly latitudes. It's a time when the days are short, the nights are long and they're both typically cold. Fortunately, a variety of holidays bring families and friends together during these otherwise dreary months and there's usually a lot to eat. People also tend to sleep more and exercise less. Unfortunately, many of us tend to pack on additional pounds during these months, and our New Year's resolution often involves some pledge about losing weight. Equally often, the pledge goes unfulfilled, so people's weight tends to creep inexorably upward—the average weight gain through adulthood is actually about a pound per year. If you've found yourself putting on this extra weight and are looking for someone to blame, you can point to the bear and his brethren.

The black bear is the most well-known mammalian hibernator, although some argue that bears really only drift into torpor. Like other hibernators such as the raccoon, woodchuck, skunk and chipmunk, the bear gorges on carbohydrate-rich foods in the late fall, gaining a substantial amount of weight in the process. At the onset of winter, the bear proceeds to find a comfortable den in a cave, hollowed out tree or rock crevice, and insulates it with leaves and twigs. The bear then rolls up into a ball, with its furred back and layers of fat acting as extra insulators, tucks its head between its forepaws and dozes off for the next several months. Eventually, its body temperature drifts down to 88 degrees

Fahrenheit (the temperature of smaller mammalian hibernators may drop to 40 degrees Fahrenheit or less).

By reducing its body temperature, the bear can cut its metabolic rate in half, which means that it can eat less, drink less and critical organs such as the heart, lungs, liver and kidney can more or less shut down. In fact, the heart rate may decrease to as low as ten beats per minute. Urine flow essentially stops, and the bear stops eating, drinking and defecating for up to a hundred days, at which point it essentially just wakes up, yawns, stretches and gets on about its business. There is pretty good reason to believe that this remarkable process has both a hormonal and, ultimately, a genetic basis. A number of different substances have been shown capable of inducing hibernation-like changes in metabolism and energy consumption in small mammals.

Birds do it, bees do it, bears do it, so why can't we? Well, we might be able to. There are any number of conceivable ways in which humans might be induced to hibernate. Since many of our mammalian relatives hibernate, it may be that we *already* have the genetic equipment to do so but have suppressed it because we have created an environment in which we can almost always find enough food—so there's no need to hibernate. Human hibernation might actually be as simple as turning on the genetic equivalent of a light switch. Alternatively, we may discover a hibernation hormone that could be trickled into an intravenous infusion for hours to months at a time, making long-distance travel much less tedious. But there are also many more immediate potential benefits from hibernation research.

One of the key features of mammalian hibernation is that the primary fuel consumed during the winter is fat, while protein-rich muscles are relatively spared, which is very different from what happens to humans who don't move for prolonged periods. Bears, for example, lose only about 20 percent of their muscle strength during their four-month winter sleep, while even with adequate nutritional intake humans lose 80 percent of their strength over a similar period. Hibernators manage this to some extent by recycling their proteins when they're dormant; resting humans discard protein into the urine in the form of ammonia. As a result, hibernators protect and preserve their muscles while at rest. A bear can wake from a four-month snooze and almost immediately charge up a hill after a fleet-footed meal, but a bedridden human after a

comparable snooze would require many months of exhausting physical reha-
bilitation to get back to a comparable baseline level of fitness.

As with the gecko's foot and the squid's beak, the lessons we learn from
the techniques used by various species to reduce metabolism and oxygen consumption in times of scarcity will eventually have a broad range of implications
for humans. For example, the current worldwide obesity trend is thought to result from a combination of changing diet patterns, decreased activity and the
human tendency to store calories as fat when food is readily available. This energy storage in the form of fat is exactly what bears and other mammals do
every fall; but we humans just never seem get rid of the fat in the spring or any
other season. Imagine if we could take lessons from hibernating animals and
develop drugs that direct the body to consume fat, or to preserve muscle. Fat
burning would obviously be useful for dieters, and muscle preservation could
help bedridden patients as well as athletes in zero gravity. The key lesson to
take from hibernation, however, is the way that animals have adapted to and
actually taken advantage of cold weather.

Animals developed hibernation as an adaptation allowing them to survive
winter's food scarcity, but it turns out that there may be other good reasons to
stay out of the midday sun and stay cool like the bear in winter. I made the
analogy between oxidation and rust in the least chapter, and many of the
processes that cause the body to figuratively rust are temperature dependent:
The higher the body's temperature, the greater the production of free radicals
(the toxic molecules that can injure normal tissue) and the greater the degree
of inflammation. Fever may be helpful in the setting of a generalized infection
but harmful when accompanying autoimmune disease or cancer. More and
more evidence suggests that we can improve outcomes after stroke and heart
attacks, improve the symptoms of certain diseases and potentially even prolong
the lives of healthy people by lowering body temperature in certain conditions.

There are several ways that cold can limit damage in diseases such as a
stroke or heart attack. The ice water baths used by elite athletes after hard
workouts prevent muscle fatigue and damage from the workout and are a useful way to understand the benefits of controlled hypothermia in disease. Put
simply, after hard work, the blood oxygen levels in muscles drop and the muscles eventually fill up with waste products such as lactic acid, which gives us

the tired and burning feeling. When an athlete eases into an ice water bath after, for example, a ten kilometer race, her thigh and calf muscles go to sleep as they cool, and the vessels and muscle fibers constrict, squeezing all of the old blood out of her legs so that it can be refreshed with new oxygen-rich blood. The cold slows down the muscle's metabolism and prevents postexercise inflammation in the hours and days after the stress: The lower the temperature, the lower the metabolism of the muscle, or the brain, the heart, the liver and kidneys. A logical question would be to ask what happens when you cool the body all the way down.

Baseball is America's national game, and it is fitting that baseball legend Ted Williams died one day after America's Fourth of July national holiday, in 2002. Williams was in failing health and had been through open-heart surgery and the insertion of a pacemaker in the two years prior to his death from a heart attack. He was the last major league baseball player with a batting average over .400, was voted the most valuable player in the American League twice and was elected to the Hall of Fame in 1966. His nickname "the Splendid Splinter" is unfortunate given what happened to his body after he died.

Shortly after his death, an argument developed between a daughter from his first marriage and a son from his third, relating to whether Williams' dead body should be cryonically preserved—that is, frozen solid in liquid nitrogen—or cremated. The son, John Henry Williams, produced a motor-oil-stained scrap of paper signed by his father, John Henry himself and a third party, indicating their mutual desire to be put into "Bios-Stasis after we die." This supposedly definitive document did not persuade the daughter, Bobby-Jo Ferrrell, who contended that it was a forgery and that Williams actually wanted to be cremated and have his ashes spread over his favorite fishing grounds in the Florida Keys. John Henry prevailed.

In controlled cryonic preservation, the newly dead body is first chilled to the temperature of ice water as quickly as possible, then cooled under dry ice until it reaches −40 degrees Fahrenheit and finally lowered into liquid nitrogen. Unfortunately, Ted Williams died in Crystal River, Florida, and had to be flown to the nearest cryonics preservation facility in Scottsdale, Arizona. By the night after his death, his body made it to an operating table where it underwent "neuroseparation surgery" in which his head and body were divided the one from the other and each was subsequently placed in a different containers. According to *Sports Illustrated* magazine, the head now rests in a liquid-nitrogen-filled steel can resembling a lobster pot and the body is in a nine-foot-tall cylindrical steel tank. Regrettably, the head has apparently cracked on several occasions—evidently a known complication of the freezing process.

Literature from the organization that now stores Williams' body, the Alcor Life Extension Foundation, suggests that the best results are obtained when blood circulation is restored almost immediately after legal death occurs, and its clients are encouraged to relocate to Scottsdale as the end approaches. Alternatively, for a fee, an Alcor crew can be deployed to a remote location and then hover while awaiting the death of one of their clients. Under these ideal circumstances, the Alcor team begins what they describe as "life support measures" immediately after legal death: immersing the body in ice water and then commencing artificial respiration and circulation using a device known as a thumper, which is an electrically powered chest compression machine originally designed for mechanical cardiopulmonary resuscitation in potentially revivable cardiac arrest victims.

After restoring breathing and blood flow, the Alcor team infuses the client's body with a cocktail of chemical agents—many of which you've probably never heard of—including free radical inhibitors, nitric oxide synthase inhibitors, poly ADP-ribose polymerase inhibitors, excitotoxicity inhibitors, anticoagulants, pressors, pH buffers and anesthetics. You could certainly be forgiven for thinking this brew sounds just like the magical brew served up by the three witches in *Macbeth*, but there is a scientific basis for the administration of each these drugs.

Free radicals damage other normal molecules, such as those in cell membranes, and potentially cause cell injury or death. Nitric oxide, poly ADP-ribose polymerase, and amino acids such as glutamate can play a role in brain damage during or after a stroke. Pressors increase blood pressure in critically ill or anesthetized patients. Anticoagulants prevent clotting. And the administration of an anesthetic to a dead man might seem like closing the barn door after the horse has bolted were it not for the fact that anesthetic drugs can actually reduce the oxygen requirement of brain cells.

The idea behind giving these drugs after breathing has stopped is obviously to prevent further brain cell damage and to preserve at least the structure of the no-longer functional brain machinery in hopes that the technologies of the future, like nanotechnology, might one day be able to restart it. The next step in the process has the same goal. Tubes are placed in the major arteries and veins, and the body is then placed on a heart-lung bypass machine exactly like the one used in heart surgery. The client's blood is removed and replaced with a special organ preservation solution and his cadaver is then cooled to a temperature a few degrees above the freezing point of water.

Depending on which plan the client signed up for with Alcor, one of two things happens next. Either the whole body is frozen or the head is removed and frozen alone. The second plan is less expensive but may be justified based on the premise that the brain is the seat of the soul. In the latter instance, a cryoprotective solution is then infused into all of the blood vessels to the head, including both carotid and vertebral arteries, at a temperature just above freezing.

During the head-preparation phase, small holes are drilled into the skull to visually monitor the behavior of the brain as it is preserved—injured brains swell and bulge up through the bone hole, and brains that shrink away from the hole are behaving more normally. Alcor's ultimate goal is what they call vitrification, which is a "stable, ice-free state," at a temperature of 320 degrees below zero Fahrenheit. At this point, the head is essentially a crystalline bust of its original owner, wherein the constituent atoms are almost completely at rest—bizarrely beautiful when you think about it.

Alcor's sales pitch is optimistic. They are selling the idea that science will eventually figure out a way to do something restorative with your frozen brain, and that even if a skull cracks during the brain freeze, all the king's horses and all the king's men *will* be able to put it together again. Admittedly, most mainstream scientists think this is hokum and that the laws of physics preclude the development of a molecular reassembly machine. Irrespective of whether or not a single Alcor client is ever reanimated, cold has a very real and increasingly prominent role in today's practice of medicine; and in all likelihood, some cryonically based process will eventually permit real, long-term, suspended animation.

In March 1975, as reported in a later *Time* magazine piece, 18-year-old Brian Cunningham was pulled from his submerged car, which had plunged through the ice of a frozen pond in Michigan 38 minutes earlier. Given the length of time under water, his undoubtedly blue color, the lack of a heartbeat and respirations, Brian was declared dead at the scene and dispatched to the morgue when he startled everyone in the coroner's wagon by belching. They diverted to an emergency room, and after two hours of CPR and a night on a ventilator, Brian regained consciousness and went on to finish college with top grades. Brian's seemingly miraculous recovery captured the popular imagination, but the underlying physiological reflexes that kept him alive are well understood.

As soon as his face was covered with the cold pond water, Brian's heart rate would have slowed dramatically, and blood flow would have been diverted away from the skin and muscle to the heart and brain. In effect, to protect the most critical organs, the body has a built-in set of reflexes that may have their origins in the birth process and that we share with diving mammals such as the seal and porpoise. For most normal people, the heart rate will slow significantly if they hold their breath and plunge their faces in ice-cold water. This so-called diving reflex would have bought a little time for Brian, perhaps a few extra minutes, but the slowed metabolism resulting from cold water submersion protects the heart, brain and all of the other organs exactly the way it does in the hibernating bear.

Normal neurological outcomes have been reported in many people who were technically dead on discovery, with no heartbeat and fixed, dilated pupils, but who were very hypothermic. Some were mountaineers buried under an avalanche, some drowning victims and some attempted suicides who wandered off into the cold. In fact, as a result of many miracle stories like that of Brian Cunningham, paramedics and physicians will usually start cardiopulmonary resuscitation even in someone who appears to be dead if they're found in cold water or snow, and continue it until they have attempted to warm them up and restart the heart.

Doctors have used intentional cooling as an adjunct to medical treatment for thousands of years. In fact, total body cooling was used in the treatment of tetany thousands of years before the birth of Christ by the Hippocratic school of medicine, and Baron de Larrey packed the limbs of unfortunate French soldiers injured during Napoleon's March on Moscow in the readily available ice to render them numb and bloodless prior to amputation. Larrey was Napoleon's surgeon-general and developed, among other things, the "flying ambulance," a horse-drawn wagon carrying a medic team that collected wounded soldiers even in the heat of a battle. Larrey is said to have performed as many as 200 to 300 amputations on certain days during the ill-fated Napoleonic campaign, treating both frostbite and bullet wounds. The protective effects of cold have been employed therapeutically in cardiac surgery for over a half-century, and Dr. Charles Bailey, the innovative, Philadelphia-based heart surgeon, was also one of the pioneers in the use of intentional cooling to protect the brain during heart surgery.

A technique called deep hypothermic circulatory arrest is used today to put patients into a kind of cold hibernation during repairs to the aorta. The aorta is, of course, the major artery for the body and blood flows through it to the brain as well as to every other organ in the body. In order to work on the portion of the aorta closest to the heart and brain, it is often necessary to shut off the entire blood supply temporarily, much as one would shut off a water main before doing major plumbing work. Hypothermic circulatory arrest was developed to allow surgical teams to replace critical portions of the vessel while the brain is "hibernating" at temperatures near freezing. We use this technique in the treatment of an often-fatal disease called aortic dissection that typically comes out of the blue and in which the surgical management makes the difference between life and death. Aortic dissection usually strikes men in their mid-fifties with high blood pressure, and the first symptom they typically report is severe chest pain, often described as "tearing." The pain the patient feels is actually the aorta ripping apart as its inner layer delaminates from the outer layer, and it is the only warning he and his doctors get before the aorta ruptures—at death.

Jack and Ellie Addison lived in a small town in eastern Pennsylvania. They had generally stayed away from doctors and medicines; although Jack had once been told that he had high blood pressure, he hadn't paid much attention to it. He was a plumber and, in his early seventies, still worked. Jack awakened early one morning around four A.M. with severe back pain. He complained rarely although he had arthritis, but this morning he told Ellie that he felt like something was tearing apart inside him. He was pale and had drenched his pajamas in sweat.

Many cardiovascular calamities strike at night or very early in the morning when blood pressure increases during the rapid-eye-movement stage of sleep. Unlike the pain of angina that can mimic heartburn or muscle pain, an aortic dissection is characterized by tearing back pain. Ellie drove Jack to the local clinic that morning, and the clinic physician's first thought was a dissection, so he arranged for immediate transfer by ambulance to the nearest real hospital, where a CAT scan showed a "probable ascending aortic dissection."

The community hospital arranged to transfer him urgently by helicopter to our university hospital for a surgical procedure by our highly specialized aortic surgical team. Within minutes of the request for transfer, the medical helicopter crew launched and headed out to pick him up. After arriving at the hospital, the crew moved quickly from the helipad to the emergency ward. Addison was bundled onto a special stretcher that was wheeled out to and locked to the floor of the helicopter. The helicopter crew wears flame-retardant jumpsuits with lots of pockets, helmets with dark pull-down visors and microphones. Addison could see the crew's lips moving as they talked into their mikes, but he couldn't hear them over the engine noise and couldn't see their eyes through the visors.

While the helicopter was picking him up, the aortic surgical team was notified of the impending arrival of a critical dissection, which set off a controlled chain reaction in an operating room as a team prepared for emergency surgery. Some of the team members were called in from home. As soon as the helicopter landed on the rooftop helipad, Jack was unloaded onto an elevator, taken down several floors and wheeled directly into the operating room. Several people immediately surrounded him and moved him from the stretcher onto the narrow, cold operating bed. They all wore surgical masks, hats and scrub-suits.

A junior surgeon explained quickly and calmly that this was an operating room, that Jack needed an operation, that it had risks and would he sign this consent form. While that doctor talked, others stretched Addison's arms out in a crucifix position onto arm supports. The people around him moved quickly and purposefully. He could hear them talking to one another but couldn't understand much of what they were saying because it was very technical. He couldn't see their lips move behind the masks, just busy darting glances at him, at monitors, at each other. A couple of them paused briefly to smile with their eyes and to squeeze his hand.

When caring for patients like Addison, our priority is to get the patient to sleep as quickly as possible without letting his blood pressure or heart rate get so high that the damaged aorta ruptures in the process. Just about everything that had happened since Addison woke up that morning might have been designed to raise his pressure, what with doctors, ambulances and helicopter rides, and he was fortunate to have lived this far. The operating room is chilly. EKG electrodes are cold and wet; jewelry such as wedding rings, clothing and the dignity of emergency surgery patients are sometimes unceremoniously stripped or, in some cases, cut off in the rush to save their lives.

At about the time that Jack Addison's wife, Ellie, drove onto the exit ramp from the highway toward the hospital, the contents of two syringes, one labeled red and the other blue, were injected into intravenous tubing in Jack's neck. He yawned once uncontrollably and then closed his eyes, his blood pressure gratifyingly stable. The drug in the blue syringe is a synthetic derivative of morphine called fentanyl, originally manufactured by chemists for use in anesthesia and now sold illegally on the street by the name of China White. The fluid in the red syringe is a synthetic drug retaining the characteristics of its forefather, curare, which is still used by South American natives on arrow tips and blow darts to paralyze prey.

Once Addison was asleep, a breathing tube was inserted into his airway; a long, bright yellow catheter was threaded through his heart; a flexible rubber tube was lubricated and inserted through his penis into the bladder; another tube was placed through his mouth and into the stomach to drain green bilious stomach contents. After that, an ultrasound probe, wrapped in slick black rubber, was slid down the throat into the esophagus to create grainy, black and white, real-time pictures of the moving structures in the chest. The pictures showed the tear in the aorta, which would ordinarily look like a simple circle but in this case looked like the Chinese Yin-Yang symbol, exhibiting the ominous separation between the inner and outer walls of the vessel. More tubes were inserted, including temperature sensors in the rectum and the nose, and electrodes were pasted on his arms, legs and head to monitor the heart, spinal cord and brain. Ordinarily when the electrical activity in any of these monitors stops, it is a sign of death, and the heart and brain will turn silent and still during this operation.

When all is said and done, twenty or thirty different monitors were inserted into or attached to Addison's now sleeping body and it took an hour before the first incision was made. With the preparations completed, the surgeons made a vertical incision in the center of his chest. There was no bleeding at first, as if the skin was caught by surprise, but blood welled up quickly. Smoky haze drifted over the operating table for the next several minutes as bleeding vessels were cauterized. The haze has a distinctive, cooked-meat smell. The breastbone was then cut open using a pneumatic saw, exposing the bruised and misshapen aorta. The large blood vessel was swollen to twice its usual garden-hose diameter, and dull red in color rather than the usual yellow-white. It expanded threateningly with each pulse. The surgeons touched it only very gingerly with gloved fingers. A stainless steel retractor was used to spread open his chest. Each crank on the retractor forces the chest open another bit. Sky-blue drapes were placed rectilinearly around the wound, framing the sick, red aorta.

The surgeons inserted still more catheters into Addison's already well-plumbed body. One was inserted into the vena cava, the cavern through which blood flows from the body into the heart; another was threaded into a major artery in his groin. These catheters are used to reroute blood circulation during the operation. A large dose of heparin was injected to eliminate clotting as the blood circulates through the plastic tubing of the heart–lung bypass machine. Pharmaceutical heparin comes from the intestines of cows and pigs.

Our aortic surgical team consists of anesthesiologists, surgeons, neurological specialists, nurses and perfusionists, each with their own set of critical tasks. At the head of the bed, above and outside the sterile field where the surgery occurs, mechanical equipment and drugs surround the team of anesthesiologists. The perfusionist is a highly trained technician who runs the cardiopulmonary bypass pump that takes over the functions of the heart and lungs during surgery and consists of many clear plastic tubes of varying thickness. When blood is rerouted through the bypass machine, the tubes fill. Initially depleted of oxygen and dark purple in color as it flows over the side of the operating table from the patient into the pump, the blood is bright red as it returns fully oxygenated back to the patient—the magic of blood made visible. A neurologist sits in a corner and monitors two computer screens showing spiky rows of brain waves and nerve conduction patterns that are intelligible only to another neurologist. They represent electrical noise from billions of nerve cells in millions of conversational clusters, as if from some other-dimensional cocktail party.

Before Addison's aorta can be repaired, his body must be cooled to 15 degrees Centigrade or about 60 degrees Fahrenheit, which is the temperature inside a morgue refrigerator. All of Jack's blood, five liters worth, is then siphoned into a large clear plastic canister, which we call the reservoir. This chilled, bloodless state is termed "circ [for circulatory] arrest," which is a euphemistic way of saying that the patient's body is cold and pale on the OR table while his blood is next to him in a jug on the floor. Red blood cells are designed to race ceaselessly, through an endless network of blood vessels, squeezing their way through capillaries and dashing through the larger vessels, for their entire 120-day life span. While each trip starts at the heart, no circuit is ever the same. These action-oriented red cells will now sit idly for an hour, drifting slowly in the reservoir and the cardiopulmonary bypass pump is now turned off.

Addison's aorta is now empty, as are all of his blood vessels, and his head is visible above the drapes outside the sterile field. It is cold to the touch, cadaver white, serene. His heart is flaccid, empty and still in the center of the sterile field, bathed in an ice-water bath. The conversational clusters on the neurologist's screens have fallen silent. The flow of urine ceases. Bleeding surfaces stop bleeding. All of the lines on all of the monitors are flat. Jack Addison is drained, cold and by almost any definition, dead. And the critical part of the operation can now begin without risk to the brain.

The surgeon cuts and removes several inches of the diseased aorta. Looking into it, one can see the entrance of the carotid artery, which supplies blood to the brain, as a tunnel heading north from the open aorta. An accordioned sleeve of bone-white Dacron graft, looking a little like the timeless children's Slinky toy, is used to replace the diseased aorta. The surgeon sews the graft to the remaining aorta with blue polypropylene suture. The knots are tied with many throws, because one loose knot can kill a patient. When the repair is completed, the cardiopulmonary bypass machine is turned back on and blood begins to flow again. Red cells that have been sitting idle in the reservoir for an hour are heated, oxygenated and reinfused into Addison's body. His head visibly pinkens and palpably warms. The urine starts to trickle again and cut tissue bleeds once more. As the warmed blood bathes and reawakens the sleeping neurons in the brain, spiky waves reappear on the neurologist's monitors.

Finally, a clamp is removed from the repaired aorta and blood flow returns to the still flaccid heart. The first visible indication of life is a slow writhing movement of the heart muscle, which we call *vermiform* or wormlike movement. The anesthesiologist's screen shows the disorganized spiky trace of cardiac fibrillation; the surgeon applies a pair of sterile, stainless spatulas to the sides of the heart and defibrillates it with a little jolt of electricity. The heart jumps in response to the charge and the wormlike movement stops. The surgeon taps the heart with a metal instrument and it contracts once, then fitfully, and finally begins to beat regularly.

When Addison is finally warm enough to be safely removed from the cardiopulmonary bypass machine, and the heart is strong enough to handle the circulation on its own, the pump is turned off. The surgeons pull the breastbone together with metal wires that are twisted tight with sterile pliers. They close the skin with surgical staples. The scar will eventually look like a railroad track. Finally, Addison is lifted from the OR table onto a hospital bed made up with clean white sheets and rolled to the intensive care unit.

It's almost ten o'clock that night before Ellie is allowed into the intensive care room. Jack has begun to waken, and the nurses think he will be comforted by her presence. Monitoring equipment surrounds him and he is still on a respirator; and like many family members, Ellie is intimidated by the equipment. She's clearly uncomfortable touching him on his hands or head for fear something will break, so she rubs his feet instead. Jack's eyes open sluggishly at first, wander across the ceiling and around the room until he sees her, and then they lock on. He smiles slowly around the tube in his mouth, which is removed shortly thereafter. He speaks normally and moves his hand to take hers.

We perform the procedure I've just described routinely at my hospital, to the point that we've almost come to expect good outcomes like Jack Addison's, but it's a miracle every time someone wakes up normally after having had no blood flowing to their brain for an hour. I once heard a scuba diver describe himself as an "air hog," by which he meant that he typically bubbled his way through a standard air tank more quickly than most of the other people on a dive and would therefore need to surface sooner. The brain is also an air hog.

The mammalian brain doesn't like oxygen deprivation, and it has evolved a whole series of mechanisms to make sure it gets enough of the life-saving gas. The first of these—and one you may have noticed if you've ever spent time in a high-altitude location—is hyperventilation: There is 40 percent less oxygen in the air at 12,000 feet than at sea level, so the brain's first response to altitude is to direct the body to breathe like mad. As inhaled oxygen concentration decreases, the brain also increases its own blood flow at the expense of muscles and other organs.

The brain uses 40 percent of its oxygen intake for cellular maintenance, and the other 60 percent in the generation of nerve impulses; in other words, it spends a little more than half of its resources on thinking and running the rest of the body. Consequently, one of the first obvious effects of oxygen deprivation is impaired thinking and clumsiness; I can speak to this personally, having participated in a medical school laboratory experiment in which we inhaled various gas mixtures. My performance on memory and cognitive tasks was perceptibly poorer while breathing a low-oxygen mix.

Hypothermia has dramatic effects on the oxygen needs of the brain. If the body's temperature is reduced from normal to 32 degrees Centigrade (90 degrees Fahrenheit), the brain's oxygen consumption rate is reduced by 30 percent. At 28 degrees Centigrade, it is only 50 percent of normal, and at 15 degrees, the temperature Jack Addison was cooled to during his surgery, oxygen consumption is down to 24 percent of baseline. Reducing the brain's temperature from normal to 15 degrees Centigrade extends the safe period for cardiac arrest from five minutes to over a half an hour.

Controlled hypothermia or medical cooling has been used for a long time in surgery, but we are now beginning to use it in the management of patients who survived a cardiac arrest as well. The remarkable thing is that it seems to work even when the cooling isn't begun until well after a patient's heart stops. A recent *Newsweek* article described what happened to Bill Bondar, who keeled over in his driveway in a retirement community in southern New Jersey one Wednesday night in the summer of 2007. Bondar returned home from a guitar jam session around 10:30 P.M. He was found by his wife on the ground several minutes later with what she described as "the eyes of a dead man." She started CPR but had to stop long enough to call 911, and it took a few minutes for the ambulance crew to arrive so there were probably long intervals during which no blood was flowing to the brain on that warm summer night. His heart was restarted with a couple of cardiac shocks, but Bondar remained in a coma; and his wife, Monica, quickly decided to transfer him to us at the Hospital of the University of Pennsylvania.

Bondar arrived here at 1:30 in the morning—three hours after the event, and our emergency room doctors immediately started a new total-body cooling protocol that involved the administration of two liters of very cold saline into the bloodstream. His body was simultaneously wrapped with a cooling blanket regulated to maintain his temperature at 92 degrees Fahrenheit. He was cooled to this level for the next 24 hours and then allowed to return to normal temperature. Bondar remained in a coma through the weekend, and his exhausted wife was finally persuaded to go home to get some sleep on Sunday night. Monday morning she was awakened by a phone call from the hospital, and her first thought was that things had gotten worse, if that was even possible.

Miraculously, Bondar had opened his eyes and was responding. His first words after the breathing tube was removed were "How did I get here." He had lost any memory from two days prior to his cardiac arrest to the moment he had awakened. The culprit responsible for his heart attack was subsequently found to be a critical narrowing in one of his coronary arteries, which was easily handled by placement of a stent to reopen the vessel.

While it might not seem to make sense to cool a patient down *after* the heart has already stopped and been successfully restarted, this treatment, like many ground-breaking medical interventions, has its roots in astute observations made on untreated patients. There are an infinite number of ways that the human brain can be deprived of adequate blood flow for periods far in excess of the vulnerable four minutes, including a routine cardiac arrest, carbon monoxide poisoning, drowning, drug overdose or strangulation. Over the years since the development of cardiopulmonary resuscitation and intensive care units, a series of odd cases appeared in the medical literature describing what seem like particularly cruel perversions of an already horrible situation—they were lumped under the seemingly banal descriptor "delayed postanoxic encephalopathy."

Delayed postanoxic encephalopathy or leukoencephalopathy is diagnosed when a patient initially appears to recover normal neurological function after having been deprived of oxygen for many minutes, but later proceeds to deteriorate dramatically, developing delirium and movement problems. It's as if the patient was given a miraculous, but time-limited, Faustian reprieve from death. The outcome seems doubly cruel for the survivor's family members who experience life snatched from death, followed by "un-death" snatched from life. A typical patient might appear to recover completely after a drowning, for example, and be fine for days, but later deteriorates abruptly.

Some delayed postanoxic encephalopathy patients eventually improve; others develop an irreversible coma or die. The precise mechanism for this rare process isn't exactly clear, but in all probability it is due to a delayed, free-radical mediated, autoimmune reaction triggered by brain cell damage, and the immune reaction begins only *after* the oxygen supply to the brain is restored and takes days to become evident. In effect, this theory holds that the immune system begins to react against the injured brain cells, suddenly perceiving them as the non-self. Somehow, controlled hypothermia, even started hours after an anoxic injury, seems to prevent the free-radical mediated injury and protect the brain as it recovers from the initial period of oxygen deprivation.

The use of hypothermia in surgery and after cardiac arrest is a therapeutic medical intervention used in the treatment of sick patients. Both NASA and the European Space Agency have begun research on cold and hibernation in cells and animals with a clear eye to the possibility of inducing human hibernation for long-term space travel. Hibernation, its warm-weather equivalent called estivation, torpor and diapause are different forms of energy conservation adopted by various species to match their metabolisms to the available energy supplies. Bees, snails, snakes and earthworms estivate in hot arid weather. Hummingbirds and frogs are some of the many animals that dial down their heart rate and metabolism cyclically in periods of torpor. Insects have periods of diapuse during which they enter a period of dormancy. The feature common to each method is the reduced requirement for oxygen and food, allowing the organism to get by with less.

The army would like to be able to put injured soldiers into some form of hibernation while transporting them from the battlefield to a surgical venue for definitive treatment. Reducing bodily oxygen consumption might be an ideal way to protect critical organs such as the kidneys, liver and brain of a severely injured, bleeding patient. Similarly, it would be advantageous to induce some form of hibernation in patients with heart, lung or liver failure, whose own sick organs are barely meeting the metabolic needs of the body, while they await a transplant. Perhaps an individual organ could be induced to hibernate while undergoing transportation from a donor to a far-distant recipient patient. We currently chill transplant organs during transport and use various chemicals to decrease the organ's metabolism, but if it were possible to freeze the metabolism of an organism, we could extend the shelf-life of donor organs and permit more transplants to occur. However, the most exciting applications of intentional hypothermia may come in the future of interplanetary travel.

Amazon founder Jeff Bezos, Google Incorporated, entrepreneur Richard Branson of Virgin Airlines, as well as principal players from Pay-Pal, Microsoft and the computer gaming industry, have all entered the space race. So have several sovereign nations. Suborbital and orbital tourism flights are now imminent. Stephen Hawking, who has already experienced zero gravity on a modified Boeing 727, is booked on one of the initial space flights of Branson's Virgin Galactic SpaceShipTwo, as are Branson himself, two of his children and his elderly parents. Space clearly represents the new frontier, and while various players have different agendas, colonization of the moon and Mars are obvious next steps.

The outward-bound limb of the *Apollo* trips to the moon in the late 1960s and early 1970s took three to four days and the astronauts had plenty to do during the trip—so there isn't any real need for hibernation during a trip to a lunar station. In contrast, a manned mission to Mars would take many months using current propulsion technology, and one of the major design considerations for the spaceship will relate to how much food and oxygen it can carry. If an astronaut could be induced to hibernate like a bear—she'd need to gain a fair amount of weight in the form of fat prior to launch date as bears gain up to

30 pounds a week during the fall—but she could then go largely without food or water and require less oxygen during a substantial portion of the trip. The interior of the ship would almost certainly be kept quite cold, which wouldn't be a problem in outer space. And all of this would make for a much more compact vehicle. Of course, we'd have to understand how to induce hibernation, how to reverse it as the spaceship neared Mars and how to monitor the astronaut during hibernation.

The real kicker about a manned mission to Mars using current technology is that it would likely be a one-way trip. While we know how to get there and how to land on the planet, we'd have a very hard time bringing enough fuel to get off the surface and back home again according to Buzz Aldrin, a former moonwalker. However, once you take this long view and acknowledge that space colonizers will be saying "Good-bye" rather than "See you later" when they leave, experimentation with reanimation technologies like the ones used at Alcor begin to seem less bizarre.

The Alcor preservation technique is designed to vitrify the client, to literally turn a human, or a human head if you go with the value plan, into glass. And the thousand-odd clients who've enrolled with Alcor believe they are frontiersmen who will be reanimated by nanotechnologists at some future date. It's possible, of course, that they'll end up like some of the more than 200 frontiersmen and women who have died climbing Mount Everest, not to mention the many more who suffered frostbite and consequent amputation of portions of arms, legs, ears and noses just like Napoleon's soldiers. The villain in frostbite is ice—the water inside frozen cells turns to ice and the ice crystals then irreparably rupture critical cell membranes. As Dr. Michael Shermer, an Alcor critic, wrote in "Nano Nonsense and Cryonics," a September 2001 article in *Scientific American:* "When defrosted, all the intracellular goo oozes out, turning your strawberries into runny mush. This is your brain on cryonics." Alcor clients are betting that nanotechnologists in the future will be able to use their vitrified head as a sort of three-dimensional blueprint to guide the construction of a new version, or perhaps that they'll be able to rehab the old one. Nature, however, may have a better solution.

The Canadian wood frog turns into a frozen frogsicle every winter—its heart and brain cease functioning and its metabolism slows to a trickle at body temperatures between 20 and 30 degrees Fahrenheit, the point at which icicles ordinarily form. The frog gets away with this by manufacturing enough sugar to lower the freezing point of the tissue; the sugar acts like antifreeze in a car's radiator. When the ambient temperature warms up, the frog thaws from the inside out, with the heart restarting first, and after a few hours the little guy is able to hop off and procreate. Similarly, the winter flounder and various plants use different strategies, but they've all figured out a way to prevent the formation of ice crystals. In fact, the gene that the flounder uses to manufacture its internal antifreeze has been inserted into the strawberry genome to create frost-resistant varieties.

A little over a year before his biological death, Ted Williams successfully underwent open-heart surgery, during which his body temperature was probably lowered to protect his brain and other vital organs. This combination of induced hypothermia and cardiopulmonary bypass as adjuncts to surgery is only about fifty years old; more than a half-million patients now undergo open-heart surgery every year in the United States alone, and many of them are able to return to their normal lives within days. The deep hypothermic circulatory arrest used during the repair of Jack Addison's aorta, when the brain was cooled to the point that it was electrically silent, has only been used routinely in adults for about twenty years. In the last ten years we have seen significant advances in the length of time we can preserve single organs for transplantation using combinations of cold preservation and molecular preservatives.

While it's very unlikely that the legendary Ted Williams, whose other nickname was The Kid, will ever be recreated from the frozen remains suspended in a tank of liquid nitrogen in Scottsdale, cold will almost certainly play an ever-growing role in the treatment of sick patients, transplantation and the expansion of human horizons. It's even conceivable that the first human space travelers to 80 Ursae Majoris, a star in the Great Bear constellation also known as Alcor, will be retroengineered with stem cells carrying genes that make antifreeze, like the wood frog, and therefore will travel in a state of cold, suspended animation.

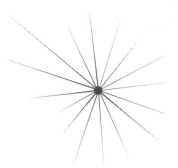

CHAPTER 11

PINK OR BLUE

In the summer of 1979, about a month before I started medical school, some thoughtful person saw fit to invite a group of the members of the matriculating University of Pennsylvania medical school class of 1983 to a professional baseball game so that we could all get to know one another before classes began. It was a very nice idea and actually turned out to be a pleasant evening, but one that I remember for the striking piece of advice given to me by one of my soon-to-be classmates, who was slightly intoxicated by the seventh-inning stretch. She was a bookish, bespectacled brunette who let us all know almost immediately that she'd already been working at medical school for the past year and was therefore "experienced." She pulled me aside toward the end of the party, leaned in close, and slurred aromatically: "Just . . . be sure . . . to . . . keep your body moist." And then she gave me a knowing, owlish wink.

At that time in my life, I fancied myself to be a pretty experienced man of the world, having taking a couple of years off between college and medical school to knock around a bit. Yet her suggestion brought me up short and left me pretty much speechless. Each of the alternative interpretations that immediately crossed my mind was equally distressing. The first was that, despite the heat and humidity of the summer, some portion of my visible epidermis was flaking off or cracking in an unappealing fashion, or perhaps there was some penumbra of dandruff floating around my head. I took a quick inventory and found none. The thought that immediately followed was that maybe this was some unusual, kinky, medical student pick-up line that I couldn't immediately decipher.

I can't remember how we got from those unpalatable possibilities to the one that she really had in mind, but I'm sure *I* arrived there with the relief of a drowning man thrown unexpectedly onto the beach, albeit mixed with more than a splash of repugnance once I understood her point. It turned out that she was actually trying to give me a little head start on cadaver maintenance, and that I was supposed to mist the corpse between anatomical dissection sessions so as to keep the subject of my dissection from drying out.

Gross anatomy was one of the very first classes in medical school, and after a group lecture by an elderly, curmudgeonly Scotsman who had once been a surgeon and eventually retired from practice to teach, all of the medical students were parceled off into groups of four, by order of surname, and each group was assigned to a cadaver. We had all been warned that it was typical for medical students to have more of an emotional reaction both to the display and dissection of certain parts than to others, and that we were therefore to start with the neck, leaving the hands, genitalia and head covered for the time being. Of course, no one paid any attention to this, and we all took a surreptitious glance at the full monty right off the bat. About half of the bodies were male and half female, so we'd all eventually get a whack at a variety of private parts when that time came.

My group was to dissect the body of a heavy-set older woman that some wag promptly nicknamed Baba Yaga, thereby demonstrating a precocious grasp of the gallows humor found in most of the older professions. Gross anatomy lab is but the first of a series of molding experiences in which young doctors learn that some part of what they've signed up for involves the violation of boundaries of propriety held sacrosanct by the rest of society. Our first incision into the cadaver of someone who looked for all the world like a grandmother was a transitional moment. A particularly memorable comment came from one, clearly shell-shocked, nerdy guy we all eventually came to like quite a bit, and henceforth nicknamed The Stallion, who stammered "wouldn't you know that would be the first time I'd see *that*," when we undraped Baba Yaga's genitalia.

I had spent some of my idle hours in the waning days of the summer of 1979 trying to imagine what medical school would be like and, of course, I got a lot of it wrong. It turns out, for example, that cadaver flesh is not pink and cadaver muscle is not red: They're both gray, as is everything else in an embalmed cadaver. It also turns out that, despite the warnings of my first medical school acquaintance, cadavers don't dry out, because some chemical reaction happens between body fat and formaldehyde, turning the two into an incredibly greasy film. The greasy miasma actually remains inside of one's nose and on one's clothes for days after contact; and it contaminates everything with which it comes in contact.

Pink is, however, the natural color of red blood cells and normal, living muscle, and some bacteria even stain a bright pink color when they are exposed to a diagnostic technique devised in the late 1800s by a Danish bacteriologist named Hans Christian Gram. The Gram stain technique is only one of a still-growing array of beautiful stains devised by pathologists to examine cells under a microscope. Histology class, which alternated days with gross anatomy and where we examined human tissue slides from all parts of the body, gave us a much more colorful sense of the body than the bland, uniformly gray tissues in the anatomy lab.

Muscle fibers are long, cross-hatched ladderlike cells, while the cells in glands such as the thyroid and pancreas are laid out like formal gardens in their respective organs. The liver and the kidney are extraordinarily interdigitated arrays of tubes that nature designed to clean and purify the blood and to pipe the toxins off into the urine or bowel. The brain is, at first glance, a seemingly chaotic jumble of neurons, but that critical organ actually has a remarkable, complicated anatomic logic that we're only just beginning to understand.

The variety of shapes and functions that cells take in the mature human being is astonishing when one considers the fact that they all ultimately derived from one single cell that resulted from the marriage of a sperm to an egg. The full span of every human's corporeal existence is bookended by the unheralded creation of a single-celled zygote in the pink darkness of the womb at the very beginning of life and by the sudden vacancy of a lifeless body at the other end. Words are insufficient to describe the complexity and beauty of the beings that can blossom from that first bland, transparent, globular seed. All of the equipment needed to create an individual capable of contemplating the span of his own existence, the history of time and the meaning of beauty is present in that first magical cell, the zygote.

The zygote develops into a blastocyst, the hollow ball of cells formed as the zygote repeatedly divides; the zygote and blastocyst are the forms taken by dividing human cells in the interval between fertilization and implantation on the uterine wall four or five days later. The blastocyst contains the inner cell mass, a group of cells now more familiarly described as embryonic stem cells. Stem cells just happen to carry the entire human instruction manual encoded in their DNA. Blastocyst stem cells go on, in some unimaginably complicated and beautiful way, to develop our muscles, nerves, skin, teeth, hair, heart, lungs and brain. They are the ancestor cells. Many believe that cancers develop from a population of primitive *cancer* stem cells in which ordinary cellular control mechanisms are not functioning properly. Leukemic stem cells, for example, may be abnormal mutants that reproduce indiscriminately, eventually interfering with normal body processes like clotting, immune defenses and the creation of red blood cells.

One of the earliest rotations during my medical residency was on a hospital floor dedicated to the care of adult patients with the two major forms of leukemia, acute lymphoblastic leukemia and acute myelogenous leukemia, or ALL and AML. While there are differences between the two leukemias, their onset is similar and leukemia can appear in patients of any age. Larry Simms, for example, was a freshman college student, the only child of parents who were both professionals who had married late and struggled to conceive. Their son was a bright, self-confident, fairly typical only child who had done well in high school and was a very good tennis player. He was admitted to top schools on both the East and the West coasts and elected to go to Stanford, which was close enough to his parents' home that the family could gather regularly. He was home for a meal at least one night on most weekends. Larry was good enough, even at Stanford, to vie for a spot on the varsity tennis team which practiced throughout the year. Even as a freshman, Larry kept up with, and occasionally beat, some of the lower-ranked varsity players during the September and October off-season workouts. He showed a degree of patience uncharacteristic in a younger player, running down every shot. He made a big impression on the coaches at first, but his game seemed to take an inexplicable nosedive after the Thanksgiving holiday. Larry was smaller than many of his teammates, but he made up for his small size with an unnatural quickness and a great ground game. In the first part of December, Larry found himself struggling more and more to catch his breath after long points, and his bones ached at night after practice in a way that he'd never experienced before. He chalked it up at first to a combination of insufficient sleep and the newness of college.

As the month of December passed, he discovered bruises he couldn't account for on his legs and arms, and his toothbrush was often stained pink with blood after brushing in the morning. He also had a few nosebleeds. Not wanting to alarm his excitable mother, he eventually went to the student health facility after practice one afternoon about two weeks before Christmas, where the doctor seemed to take his complaints pretty seriously. The doctor ordered a series of blood tests including a complete blood count and some x-rays.

At that time, unlike today when computers perform the task, blood counts were done manually by special laboratory technicians using a particular technique. A single drop of blood was placed on one end of a glass microscope slide with a pipette. A second slide was then placed on end across the middle of the drop and pushed across the surface of the first, smearing the blood droplet into a thin layer. The blood-smeared glass slide was then allowed to dry, forming a dull maroon surface. That smear was then stained with special blue and pink dyes and examined under the microscope. Back then, every blood count on every patient involved this same, time-consuming process, while today blood counts are done by highly automated, robotic lab systems.

When viewed under a microscope, a normal peripheral blood smear shows a uniform jumble of pinkish, doughnut-shaped red blood cells punctuated by the occasional blue-stained white blood cell, the mature form of which is slightly larger than the red corpuscle. The third normal cellular element of blood is the much smaller platelet, which is the blood-clotting cell that stains both blue and pink. Platelets look like little tugboats docked next to the comparatively enormous red and white blood cells.

The vast majority of the smears that a technologist examined every day in the manual cell-counting era were totally normal; at the end of every day, a hematology technician would go home with oceans of mostly pink and occasional blue cells swimming in front of their eyes. All day long they'd gone through the same monotonous process slide by slide: Mount slide on microscope's stage, rotate low-power objective lens into position, scan blood for general characteristics, put drop of oil on smear, rotate high-power lens into position, count number of cells per high-power field, record data, next slide. The microscopes weren't much more advanced than the ones used by Leeuwenhoek, and the technicians used one of those hand-held tally counters that go up to 999 for cell counting. Every once in a while, however, the technician could see that something bad was going on at the first glance through the low-power lens, because the whole smear had a bluish tint. Larry's smear was one of those.

The most obviously unusual feature of his blood was the predominance of blue-stained cells, lymphocytes in this case rather than monocytes or granulocytes, which are the other types of white blood cells. Larry's total lymphocyte count was a hundred-fold higher than normal; and the lymphocytes themselves were bizarrely shaped simulacra of normal lymphocytes. There were also very few platelets and a lower than normal red blood cell count, which accounted for his bruising, bleeding and fatigue. One glance at the smear was all that the technician needed to make the diagnosis. Larry had acute lymphoblastic leukemia and was, at that point, only days away from death due to an infection or bleeding if left untreated. Even now, he had a fever and was bleeding under his conjunctiva (the white part) of his eyes.

The student health doctor got a phone call from the hematology lab informing him that the smear was markedly abnormal and giving him a provisional diagnosis. He told Larry that there was something abnormal in his blood that required his admission to the hospital. While the doctor didn't put a name to the diagnosis, Larry was a perceptive kid and could see the concern in the doctor's eyes. He was terrified and immediately called his parents, who had always been able to fix things for him in the past. But this time, no fixing was feasible other than to make sure he was in the best possible hands for treatment.

It would quickly became apparent to Simms's parents that their son's illness was a problem unlike anything they'd experienced before in their largely successful, happy lives—it couldn't be managed by force of will or its surrogate, money. They arrived on the scene within hours and helped to get him admitted to the hospital, still in his tennis clothes, where he underwent the usual series of tests including a bone marrow sample, spinal tap, more blood tests and x-rays. Within a matter of hours, Larry was embarked on the first phase of a treatment that was designed to kill off as much of the abnormal population of lymphoblasts (typically members of a single clonal family of cells) as quickly as possible without killing the patient in the process.

The next eighteen days of Larry Simms's life could have been drawn directly from the Old Testament. If there *were* a God at work, his interventions would seem capricious and arbitrary. Cellular famine in the midst of plenty, pestilences, exterminations and plagues all took the stage at one point or another. His bone marrow proved to be packed with abnormal cells, as is characteristic of leukemia, and the prescribed chemotherapy would need to kill off all of the pestilential cancer cells in the blood and the marrow in order to be effective. In turn, the absence of functioning white blood cells let loose the dogs of infection and bleeding.

The use of toxic chemotherapeutic chemicals began within hours of Larry's arrival in the hospital. He began taking intravenous fluids at the same time, because the first thing that happens at the beginning of treatment is a massive die-off of the leukemia and the cell carcasses need to be washed from the system into the urine. Within two days, Larry's blood smear went from a bluish-pink to an even more ominous pure pink devoid of blue, as his white blood cell count dropped to near zero. His infection-fighting capacity went with it, and his platelet count dropped precipitously. Where there had previously been bruises, there were now blood blisters all over his skin. Chemotherapy for leukemia wipes out most of the marrow's normal blood production cells, or blasts, as well as their leukemic cousins. Since blasts ordinarily act like queen bees, continuously replenishing normal cells that wear out and die off, all blood production essentially came to a stop pretty quickly after Larry began treatment.

The mother of all blood cells is called a hemocytoblast, and it's capable of differentiating into the red blood cell precursor, called a proerythroblast, as well as into the myeloblast, lymphoblast, monoblast or magakaryoblast that are the precursor cells for granulocytes, lymphocytes, monocytes and platelets respectively. In those days, we intentionally bombed out the bone marrow with chemotherapy in hopes that some heroic, hardy blast cells would emerge from the smoke and flames, with their metaphoric faces covered in grime and etched with fatigue but still smiling and ready to get back to their jobs. Above all, we hoped that those survivors would be the good guys, not the bad guys.

Usually we—the aggregated doctors, nurses, friends and families—sat there watching anxiously for seven to ten days, checking surrogate reports from the front (blood smears from the marrow) before we got any sense of which way the battle between good and evil cells had gone. We pumped in occasional fresh recruits in the form of transfusions of red blood cells and platelets; and almost invariably we ended up giving a lot of antibiotics, because the jackals and vultures of cancer treatment are the ever-vigilant, ever-hungry bacteria, viruses and funguses that prey upon the weakened human body.

Typically, by the tenth day after the start of treatment we usually started to see some welcome signs of marrow regeneration—some signs that the flood had crested, that Noah's ark had touched on Mount Ararat and that the animal pairs were out and about. Fourteen days into Larry's course, however, he remained seriously ill with some unclear systemic infection, and there was no sign of land. On the fifteenth day, we decided to take an admittedly impatient peek at the bone marrow by doing a biopsy to see what was actually going on. This was risky, because by so doing, we could precipitate a lot of bleeding.

The marrow sample we had taken two weeks earlier, before starting chemotherapy, had been crammed with abnormal, dysfunctional lymphoid blast cells, all stained a threatening dark blue. In fact, it was almost impossible at that point to see any normal marrow cells. It was as if Larry's marrow had gotten some garbled message from cell-control central command mistakenly putting it on a wartime footing—and that it had consequently been completely retooled to produce huge numbers of some product completely irrelevant to the conduct of war, such as toilet seats.

The marrow sample we saw on day fifteen, however, was strikingly different—it was a barren, post-apocalyptic wasteland. There were some spikes of calcium, some clotted red cells, but none of the hoped-for queen blast cells that form new blood. In fact, the marrow beehive was essentially vacant, and there was no evidence of marrow regeneration.

This was the point at which we began, as medical practitioners put it, to hang crepe with the Simms family. They had taken a hotel room next to the hospital so that one of them could be at Larry's bedside round the clock, and teammates and friends were conducting a more or less continuous vigil in the hallways outside the ICU. But no visitors other than close family were allowed into the room, and only then in protective clothing to avoid the possibility that visitors might bring in some even more virulent infection than the ones that were already tearing their way through his system.

Two months earlier, Larry had been a tan, muscular, self-assured 150-pound youth, confidently walking the campus of one of the greatest universities in the world as one of its best and brightest. His parents used any opportunity to tell us he was "full of life" and "a fighter," as if by so doing they could further energize our efforts. Today, however, he was fifty pounds heavier, all of it water weight, and swollen so as to be almost unrecognizable. The nurses left handprints in his skin when they moved his limbs, and the prints flattened out slowly like wet footprints on the beach. A bright red rash covered his skin, and tubes attached to machines emanated from nearly every orifice, both natural and manmade ones. The once-bright blue sparkle in his eyes had dimmed, and its only remnant could be seen in the many pictures of him with friends and family in better times pasted around the room.

Over the next two days, Larry's neurological status worsened and it became apparent that he had developed meningitis. Infectious disease consultants and neurologists came by and various alternatives were considered, but a subtle shift became evident in the communications between Larry's doctors and his parents. All of the previous discussions had invariably concluded with a "next steps" bullet point; the doctors now left a little yawning, empty space at the end of their updates, into which neither of Larry's parents had yet had the courage to leap. The nurses touched them solicitously on their hands or the back of their arms as they entered and left the room.

Early on the morning of the eighteenth day, the night nurse went through her standard neurological check and found that one of Larry's pupils no longer responded to light; it was dilated as if in fright. He was sent off immediately for an emergency CT scan that showed severe brain swelling. Within a matter of hours, his brain started to expand into the opening called the *foramen magnum* (Latin for big hole) through which the spine passes at the base of the skull, slowly crushing the brain-stem and thereby shutting off all of the critical neurological circuit breakers one by one. Shortly after 5 in the morning, with his parents standing at his bedside, Larry died.

At the start of the 1980s, five-year survival for acute lymphoblastic leukemia was about 50 percent, whereas by the latter part of that decade greater than 60 percent lived that long, and today upward of 80 percent of patients can expect to be alive five years after diagnosis. One of the critical things that doomed Larry in 1985 was the failure of his bone marrow to regenerate after the initial dose of chemotherapy. Perhaps we erred too high in the doses of chemotherapy we gave, or perhaps he was just too far along by the time he got to us, but the bottom line was that we had killed off the leukemic blast cells and there were no normal precursor cells left as survivors. Now, we have solutions. The treatments for leukemia and a variety of other diseases are much different today, not so much because we have better chemotherapy drugs, but because we have developed better exit strategies.

Today we have stem cells—cellular Adams and Eves—that can be dispatched into the bloodstream with a mission to find the bone marrow, take up residence and propagate. In fact, today we even have infusible aphrodisiacs in the form of genetically engineered molecules that stimulate stem cells to ramp up the production of normal red cells, white cells and platelets. The red cell propagation factor has even been used as a form of blood doping by athletes looking for a little more oxygen-carrying capacity in their blood. And we can well imagine what would happen if oxygen-packet nanoscale respirocytes, like the ones I described in an earlier chapter, were made available today.

Hematopoietic, or blood producing, stem cells were the first form of stem cell treatment to come into common clinical use. Stem cell bone marrow transplants were actually already being performed during the era Larry Simms died from the lack of a functioning marrow. Unfortunately, marrow transplants were regarded as highly experimental at that time and were not performed at most hospitals. The very first marrow transplant was reported in 1957 and it involved the treatment of a leukemia patient with total body irradiation to kill the malignant cells, followed by the infusion of bone marrow from his serendipitously healthy identical twin. The patient went into a complete remission, and the doctor who conceived of and performed the transplant, Dr. E. Donnall Thomas, received the Nobel Prize for Medicine in 1990.

Subsequent pioneering work was done with marrow stem cells to treat the inherited immune deficiency disease sometimes called the "bubble boy" disease. Television and movie dramatizations have been made of the life of David Vetter, who spent most of his thirteen years of life in a medically sterile environment, including the first ten years during which he never left the Texas Children's Hospital.

Stem cell transplants have now been used to treat a wide variety of diseases, and there are already international registries designed to facilitate identification and access to marrow that is compatible with a patient's own immune system, obviating the often-unfillable need for a twin or family member with closely matching tissue. Dr. Thomas and his colleague and fellow Nobel winner Dr. Joseph Murray understood in the 1980s that bone marrow contains cellular elements capable of regenerating blood and the immune system, but they didn't know that the critical elements were stem cells. As Thomas put it much later, "In retrospect, I'm surprised at how stupid we were." Fifty years after that first transplant, however, there is *still* a lot we don't know about stem cells. For example, bone marrow is transplanted into recipients through an intravenous catheter in a big vein—not directly into the bone marrow—but the transplanted marrow cells are willing to work pretty hard to find their way into the recipient's bone marrow spaces. It turns out that marrow stem cells are pretty particular about where they work, and they only begin to reproduce once they've fetched up in the right place.

Intuitively, you'd think that marrow injected into a big vein in the neck would wash around in the bloodstream forever like flotsam and jetsam, or, to be more precise, *just* jetsam. (Flotsam is the wreckage that gets into the water without anyone's active assistance, and jetsam, technically speaking, refers to goods *intentionally* cast into the ocean to lighten a ship.) You might imagine transplanted marrow cells like hemocytoblasts, erythroblasts, lymphoblasts and myeloblasts bobbling witlessly through veins and arteries like those numbered ducks in touring carnivals, eventually coming to rest in some random location, perhaps under a fingernail or at the tip of a tongue. Fortunately, nothing could be further from the truth.

Transplanted bone marrow stem cells actually seem to know exactly where they want to go, and they act like certain species of fish and birds—actively homing in on blood vessels in the recipient bone marrow. Blood stem cells recognize what are called adhesion molecules that are found only on the walls of marrow vessels; they metaphorically grab hold of those molecular hands and actually squeeze their way through the walls into the bone marrow, like celebrities through a rope line. Once in the marrow, they find a cellular ecosystem that is custom made for blood cell production.

The stem cells find the marrow environment irresistibly sexy. They *want* to settle down, and once there they feel *fecund*. Furthermore, this complicated cellular homing process is remarkably rapid and may take only a few hours from the initial transplant injection into the bloodstream, although the subsequent engraftment and cellular division typically takes days to weeks, much as it takes us humans some time to settle into a new location. But the fact that it happens at all is a testament to Nature's skills at sleight-of-hand. Like a skilled social engineer, nature seduces the transplanted stem cells into going just where she wants them to go and doing exactly what they're designed to do—to go forth and propagate.

In order to orchestrate this complicated feat, Mother Nature has actually devised a whole perfumery full of pheromones that she can deploy to seductively waft cells to desired destinations in the body. Marrow stem cells, for example, get weak at the knees in the presence of a molecule called stromal cell-derived factor–1 (the cellular equivalent of Chanel No. 5), and they amass like zombies in areas from which that molecular smell emanates. Similar chemicals are involved in a wide variety of other normal maintenance processes in the body, including the immune system's response to intruders; although in the latter instance, the pheromone may smell more like "napalm in the morning" than Chanel No. 5 at twilight. For example, a patrolling white blood cell that happens on a gang of unruly bacteria might raise its virtual tail and send out a spray of alarm pheromone that summons additional, well-armed colleagues to help corral the infection.

Bone marrow stem cells were the first stem cells we discovered and learned how to use therapeutically in medicine, and stem cell research is, of course, now undergoing explosive growth all over the world, spawning a field known as *regenerative* or *reparative medicine*. Stem cells have two distinct characteristics that make them unique: They are more or less immortal, and they can turn into highly specialized cells under the right circumstances. They are godlike cells, remaining perpetually young and yet endowed with the ability to shape-shift into the form of any of their offspring cells. The hemocytoblast, for example, is the Ur-blood stem cell and can, depending on prevailing conditions, specialize into cellular transshipment of oxygen and carbon dioxide in the form of red blood cells, perform cellular homeland defense as a white blood cell, or make cellular leaps into breaches as platelets. Similarly, the brain contains super-stem-cells that can transfigure into any of the three major brain cell types: astrocytes, oligodendrocytes or neurons.

Until recently, scientists, ethicists, politicians and even governments have drawn a sharp distinction between two types of stem cells. Hemocytoblasts are an example of what are called *adult* or *somatic* stem cells that are found in a specific body tissue or organ, such as the bone marrow or brain. Adult stem cells were originally thought to have their special abilities only when in that tissue—like fictional creatures that become gods only when they enter some magical world. *Embryonic* stem cells, on the other hand, are the mother of *all* stem cells and are usually extracted from human embryos in their undifferentiated, blastocyst stage. Using an analogy from chess, adult stem cells have historically played the bishop to the embryonic stem cell's queen. Embryonic stem cells have historically come from embryos that would otherwise have been discarded by fertility clinics. These extra embryos were successfully conceived from sperm and eggs donated by couples seeking pregnancy, frozen temporarily as blastocysts, and then, for whatever reason, no longer wanted by the parental couple. The potential repurposing of these embryonic cells that might otherwise go on to become human beings provoked ethical debate. Stem cell research proponents feel that the potential benefits of embryonic stem cell research outweigh the ethical issues inherent in tinkering with embryos; opponents feel the reverse.

One of the many reasons that we're all so interested in stem cells is that most of our organs are like shoes. The soles and heels of the brand new organs we all start with are bright and shiny and sharp-edged at first but wear after a lot of use. A given organ begins to function less effectively as it ages. As with a shoe's soles, some people go through an organ more quickly than others; some wear out the heel in a certain pattern, some the toe, depending on the way they walk through life. Drink too much and one tends to wear out the liver pretty quickly. High blood pressure wears out the heart. Too much sun ages the skin. Constant blood sugar fluctuations may wear out the pancreas and cause diabetes. And too much television or video gaming . . . well, we're not sure yet, but it definitely wears out *something* according to some parents and many childhood psychiatrists.

Highly differentiated organs, such as the liver, kidney, heart and brain, have a relatively limited ability to repair themselves or regenerate, and while you can't yet find the word *regenerativist* on the Internet, when such a job title becomes reality, these folks will specialize in extreme makeovers of your entire body, not just the trimmings that we do currently. A little squirt of something down the coronaries followed by a squeeze of stem cells and you may end up with the heart of a teenager without all the pimples. Stem cells have also been teed up to solve a lot of other problems, including genetic diseases, specific neurodegenerative diseases and such injuries as burns, heart attacks, strokes and spinal cord damage. The potential promise of these therapies is enormous, but significant barriers remain before they'll see widespread clinical application.

We know that nature has figured out a way to take the hundred or so more-or-less identical embryonic stem cells in a primitive blastocyst and eventually turn them into a you or a me. And we know that those embryonic stem cells proliferate and differentiate into all of the more than two hundred specialty human cell types along the way. We have intricately detailed etchings of every step along the road of development done by anatomists; we have beautiful intrauterine photographs; and, of course, there are billions of photographs from our increasingly well-examined lives showing the process of human growth and aging in endlessly original detail. So we know *what* Nature's doing but, as with all great magic tricks, we don't completely understand how it's done—the creation of a human from a cell—at least not yet. It's a pretty good bet, however, that we're going to keep worrying at it until we nosy humans have figured it out, *particularly* if immortality is one of the potential bonus prizes.

Until very recently, most of us belonged either to the embryonic or the adult stem cell camp, depending on whether or not we thought it was ethical to use cells from discarded human embryos for research. The general underlying presumption on both sides was that an adult stem cell, such as the hemocytoblast of the bone marrow, was merely a sort of degraded *Readers Digest Condensed* version of an embryonic stem cell. In other words, *adult* stem cells were believed to be the edited or dumbed-down versions of *embryonic* cells from which it was no longer possible to re-create the original. Recently, however, two groups of scientists appear to have shown that the developmental clock can be turned backward, and adult stem cells can be reprogrammed to revert to embryonic stem cells. In other words, it seems as if none of the original set of instructions actually get shredded—some of the material is just genetically redacted, or turned off, as embryonic stem cells devolve or evolve into adult stem cells. Each of the two research groups reported different genetic reprogramming methods and used different sources of stem cells, including facial skin, connective tissue and, curiously, foreskin from a presumably male baby. While the widespread use of pediatric foreskin as a donor source would presumably stoke another religiously divided firestorm of controversy, it appears that there are probably a variety of ways to obtain and train adult stem cells to behave like embryonic stem cells.

So let's indulge in a thought experiment and imagine you want to get into the business of making replacement organs or resoling the old ones using stem cells. In theory, all you would need to open your business is just one stem cell, since stem cells can, by definition, make more of themselves, make all other cells, and are immortal—right? So, the first thing you'd want to do with your single, precious starter cell is to make more cells so that you have a back-up supply. To do this, you'd grow your starter cell in nutrient broth and let it replicate. Eventually, you would acquire a limitless inventory of immortal stem cells that you could turn into any cell type you want, and you would be ready for your first customer. In fact, the next Noah may need only a rowboat with room for a beer cooler full of stem cells from all living animal and plant species. However, it turns out there's a rub: your customer must be a perfect immunologic match with your stem cells; if your stem cells *don't* match your customer, you've got damaged goods. Unfortunately, the likelihood that your random stem cell will match a random customer is near zero.

Stem cells, like any other tissue, carry surface protein markers, called human leukocyte antigens (HLAs) that serve as identification tags for the immune system. If a cell displays the right HLA identification credentials, the immune system assumes it is self and ignores it. But if a cell shows the wrong labels, the sirens go off and the immune system tries to eject the foreign tissue from the premises. HLA matching is performed on any transplanted tissue, stem cell or not, to make sure that the donor and the recipient are compatible. This limits the possibility that the recipient will reject a transplanted organ, or that white cells from the *donor* organ will attack the recipient. Every cell in the human body has six major HLA tags as well as several minor ones. While the probability is about one in four that siblings will HLA match one another, the odds are very low that one random individual will be a match for any other.

Interestingly, HLA markers are highly variable and can therefore be used as a fingerprint to evaluate the distribution of certain populations or tribes of people. There is, for example, an HLA type characteristic of individuals from the western part of Ireland that is highly prevalent in that area and otherwise found only in countries to which the western Irish have migrated. In fact, the British Isles as a whole show a limited set of HLA types compared to other countries, such as the United States, that have seen more broad-based immigration patterns. Tissue matching databases, such as those maintained by bone marrow registries and organ procurement organizations, have been developed to efficiently HLA-match donors to recipients. Depending on the tissue of interest, registries may be used primarily for regional or international matching. For example, solid organs (the heart, lungs, liver, pancreas and kidneys) can't be transplanted into patients too far away because organ shelf-life is limited. So you couldn't, for example, realistically bring a heart from Europe to the United States to fulfill some rare donor-recipient tissue match. Stem cells, on the other hand, can be frozen and transported freely across the world.

The traditional bread-and-butter use for stem cells as a medical treatment, as we saw earlier, has been bone marrow transplantation. While originally designed as an adjunct for the treatment of leukemia, marrow transplants are now used for a variety of cancers for which it is necessary to give as high a dose of chemotherapy or radiation as possible without permanently wiping out the vulnerable progenitor blood-forming cells. For example, bone marrow transplants are used in children to treat pediatric leukemia, severe metabolic diseases like Hurler syndrome or gargoylism, cancers of the brain and kidneys, immune deficiency diseases and even sickle cell anemia. In adults, marrow transplants have been applied to a variety of diseases in addition to leukemia, including breast cancer, multiple myeloma and other cancers.

The way a bone marrow transplant works in the treatment of cancer is somewhat different from its use in the treatment of diseases such as immune deficiency and sickle cell. Chemotherapy and radiation are used in high doses to kill off as many of the cancer cells as possible and to make room in the bone marrow before the donor bone marrow is injected into the bloodstream. Some of the stem cells from the donor will eventually make their way into the recipient's marrow and resume blood cell production over time. Unlike cancer treatment when Larry Simms developed leukemia, doctors can now treat cancer much more aggressively, secure in the knowledge that they can restore the patient's bone marrow. While the process is similar in the treatment of immune deficiency and sickle cell, the goal is slightly different: Chemotherapy is given in this case to clear malfunctioning, rather than malignant, cells from the recipient marrow and to make room for the donor's cells rather than to treat cancer.

Many medical centers have become very successful with bone marrow transplants and have good survival rates, depending on the underlying disease. Fortunately, the proliferation of bone marrow stem cell registries has made it increasingly possible to find donor sources even for people with rare HLA types. While bone marrow transplants are already commonplace, stem cell therapy may eventually be used in the treatment of a host of applications including the treatment of strokes and neurological diseases such as Parkinson's, as well as heart failure and cancer. Stem cells will also be used in test-tube drug testing and as a delivery vehicle in gene therapy.

While the clinical manifestations of Parkinson's disease and diabetes are different, they are similar in the sense that critical cells stop working in both diseases. Brain cells stop making dopamine in Parkinson's and pancreas cells stop making insulin in diabetes. We may soon be able to rejuvenate the function of these organs, much as a shoemaker resoles a pair of oxfords, by giving patients stem cells that turn into dopamine-secreting cells in the brain and insulin-producing islet cells in the pancreas. Brain stem cells may eventually be used to create new neurons in the treatment of stroke and degenerative diseases such as amyotrophic lateral sclerosis (Lou Gehrig's disease, the condition that afflicts Stephen Hawking). Stem cells have also shown promise in the treatment of spinal cord injuries from trauma, such as the neck injury experienced by Christopher (Superman) Reeve, who was an outspoken advocate for embryonic stem cell research. Although the brain and nervous system were once believed to reach a stable and fixed state of maturity in adulthood, we now know that new cells and new brain pathways are formed continuously through one's lifetime, even into old age, and that neural stem cells are responsible for that ongoing regenerative activity.

Heart failure is the common endpoint for a variety of cardiac diseases, including heart attack, high blood pressure, certain viruses, valve diseases and various toxins. Hypertension and diabetes are increasingly prevalent in the Western world for a variety of reasons, and we don't have great treatments for the patients who develop cardiac diseases as a result. Today's treatment alternatives include relatively unpleasant and expensive options such as heart transplant and mechanical heart support or replacement machines. The treatment of heart failure with stem cells that might regenerate dead or injured heart muscle is of great interest. A good deal of research has already been done on the use of stem cells to treat heart failure, but we don't really know exactly how to use the cells. Stem cells have been given, variously, into major veins, squirted down the arteries of the heart and injected directly into cardiac scar tissue. Unfortunately, under current application methods, most of the injected stem cells typically die due to physical stress, inflammation at the site of injection, insufficient oxygen or immune reactions. But stem cells are not always heroes or victims—they can also attack their recipients in certain circumstances.

One of the problems with transplants, aside from the constant threat of infection, occurs when white blood cells in transplanted donor tissue attack the recipient, which we call graft-versus-host disease, or GVHD. Immune cells in marrow acquired from a donor, particularly one with an imperfect HLA match, may attack the recipient's lungs, skin, liver, digestive tract and even the recipient's own immune system. In effect, the donor's army of white cells goes to battle against the recipient's, like marauding football hooligans, each wearing their own HLA-specific team scarves. GVHD is ordinarily treated with drugs that suppress the immune system, and severe GVHD can be fatal. However, we have discovered an interesting way to channel this ordinarily destructive complication into a potential treatment for some types of leukemia. Donor bone marrow contains what are called natural killer cells, or assassin cells that kill other cells. In certain circumstances, these killer cells can be induced to kill leukemia cells in a marrow recipient so that the recipient gets the combined benefits of a marrow transplant that kills off any bad actors remaining after chemotherapy.

The stem cells banked in HLA registries come from several different sources. Aside from embryonic stem cells, stem cells can be acquired from donor bone marrow or peripheral blood, and, increasingly from a new source— umbilical cord blood. Cord blood is the blood that remains in the placenta and umbilical vessels after a baby is delivered and the cord is cut; and it is possible to extract and bank a lifetime's worth of perfectly matched cells for a baby should they need them down the road. In fact cord stem cells can be used for non-related recipients as well.

Unlike adult stem cells and even embryonic stem cells, cord stem cells are even *less* mature immunologically. As a result, they are less likely to provoke an immune reaction from a nonrelated recipient and are also less likely to attack the patient in graft-versus-host disease. The potential promise of cord blood is so great that entrepreneurs have begun cord-blood banks as business ventures, and encourage parents to store blood harvested at the time of their baby's delivery as a kind of biological insurance vehicle should the child ever need access to some future stem cell-based treatment. Not surprisingly, of course, there are enrollment and storage fees that make cord-blood banking a possibility only for those families that can afford to pay for it. Another problem with donor and cord stem cells is that the recipient needs to be a reasonably close HLA match with the donor for any reasonable chance of success.

A desirable alternative to stem cell libraries and cord-blood banking would be the design of a sort of *universal* stem cell, one that acts like the master keys that security guards and apartment superintendents carry. A recent breakthrough has shown that it is now feasible to prospectively create a library of universal or master stem cells, which suggests that it will eventually be possible to look up the universal stem cell that fits a given recipient in the same way that a security guard looks up the master key fitting all of the doors in a specific zone of a building. At some point, we may even be able to reengineer our own adult stem cells harvested from a blood sample so that they revert to their embryonic state. We will then be able to create our own, perfectly HLA matched, embryonic stem cells on demand, so that the need for cord blood, registries or universal stem cells disappears.

In addition to their use as direct patient therapies, another area in which stem cells will have a substantial impact on medicine is in drug testing. Traditionally, new drugs are tested in three steps: Phase I in animals, Phase II in small groups of humans for safety and finally in large Phase III trials to determine whether the drug actually does what it is designed for. Not surprisingly, drugs that work in animals may not work in humans, and sometimes drugs that seem safe in animals are toxic to humans. Human stem cells will allow us to develop test-tube stem cell lines with which we can explore the safety and potentially the efficacy of new drugs before they're ever given to a human, and to do so more accurately and less expensively. And we're getting closer to the day when we will be able to test a battery of potential drugs on test tissue created from stem cells; in other words, we'll soon be able to determine the *best* drug for every *individual* patient rather than having to rely on the average results from large groups of Phase III test subjects. The use of highly individualized treatments is what is known as personalized medicine, discussed in the next chapter.

A final area of intensive current research is in the use of stem cells as a component of gene therapy to treat diseases having a basis in damaged or defective DNA, such as cystic fibrosis. The first gene therapy trials involved the transfer of functional DNA into a patient's cells using a virus as a vector or carrier vehicle. In some cases, the patient's immune system reacted violently to the viral vector. Another problem with this approach is that the new replacement DNA may get into the wrong place in the patient's own DNA. The replacement DNA may not even arrive at its intended destination if the target organ is not easily accessible. An alternative, albeit more complicated, approach is to use specially designed stem cells as the vehicle for delivering the repaired or functional DNA.

To understand how this might work, imagine the following scenario for the treatment of such diseases as cystic fibrosis, sickle cell anemia, hemophilia or muscular dystrophy in which there is usually one single defective gene. In a laboratory, stem cells are biologically engineered so that they contain a working, functional section of DNA designed to replace the defective, disease-causing DNA in the patient. The functional or repaired stem cells breed in a growth medium until there are lots of them. These brave new cells are injected into the patient, perhaps directly into the lungs by aerosol or through the bloodstream. The newly functional cells engraft, and the cystic fibrosis, hemophilia or muscular dystrophy is cured. As we become successful with the treatment of single-gene mutation diseases, we may eventually be able to treat conditions in which multiple genes are defective, such as Alzheimer's and cardiac disease.

Stem cells have already been used for a variety of novel therapies such as the growth of brand new corneas from corneal stem cells for patients blinded by damage to that thin sheet of tissue over the lens of the eye. We may soon be able to cure some of the deficits common to old age, such as tooth loss, deafness and baldness by growing new teeth, hearing cells or hair follicles. At some point farther down the road, we may become sufficiently sophisticated with stem cell manipulation that we can even grow new organs from stem cells.

All human cells derive their ancestry from stem cells, and it's possible to imagine that we humans, who have learned so much about the atom, the universe and DNA, will eventually understand how to re-create what we believe to be one of nature's greatest works—ourselves—in parts or as a whole. Right now, we're like the first human societies who discovered that they could cultivate their own food. We're experimenting with various approaches to sowing stem cells before we have a full grasp of the relevance of water, plowed soil, fertilizer and scarecrows. We don't yet know how to prepare the soil of a target organ with the right signaling and growth factors to feed and water freshly sown stem cells,; and we don't know how to ward off hungry inflammatory white cells. But consider how far human agriculture has come since our ancestors first inadvertently scattered discarded seeds on untilled land and serendipitously discovered a crop growing there sometime later.

Stem cell research is in its infancy, but the dramatic advances we've made in the years since we stood by helplessly watching leukemia patients like Larry Simms die from infections resulting from the lack of a functioning bone marrow are long past. Our blossoming understanding of the ways we can engineer and manipulate stem cells for human treatment and regeneration suggest that medicine is on the verge of a brave new world.

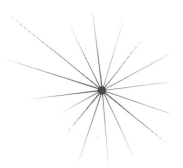

CHAPTER 12

THE NEVER-ENDING STORY

In 1209, a soldier dispatched by Pope Innocent III, one Arnaud-Amaury, laid siege to the town of Beziers in Southwestern France during the Albigensian Crusade. This particular Crusade was designed to eradicate the Cathars, who denied the existence of the Holy Trinity and held other heretical beliefs; but Beziers had a mixed population of Catholics and Cathars. While they were busy making the rounds tightening up the stays on the siege engines and cheering up the troops, one of the junior Crusaders hesitantly asked Arnaud-Amaury about how to handle the awkward problem of distinguishing between the Catholics, who were innocent bystanders, and the enemy Cathars when it came time to invade the town. The papal legate's answer has echoed through the ages. He said *"Caedite eos. Novit enim Dominus qui sunt eius,"* for which the correct translation is "Kill them all. The Lord will recognize his own," or "Let God sort them out." The succeeding massacre resulted in the death of twenty thousand men, women and children, many of whom were mutilated, blinded, dragged behind horses or used for target practice before being killed. History has not recorded what percentage of the dead was Catholic and what percentage Cathar, but it's a safe bet that large numbers of inoffensive Catholics were killed that day.

While the legate's approach seems a bit unreasonable when viewed through history's lens, his nondiscriminatory approach is similar to the one we are often forced to take in medicine: most of the traditional therapies designed to eradicate disease are relatively imprecise, and they often cause some collateral damage to normal tissue in the rest of the body. This secondary injury may come in the form of what we call side effects, like headaches and nausea, or allergies or, in some cases, actual permanent damage to organs like the kidney or the heart. Sometimes a drug that works well with limited side effects for most people, may have severe side effects or toxicities for a small subset. In other situations, doctors prescribe a treatment that they *know* will damage normal tissue because the target disease is only susceptible to toxic interventions. Chemo- and radiation therapy for the treatment of cancer typifies this problem. Too many traditional treatments are "blunt instruments"; in the future, in the era of *personalized medicine*, we will have more and more therapies that act with the precision of a laser.

Radiation was first used to treat human disease at the turn of the last century, beginning immediately after the Nobel prize–winning discoveries of x-rays by Wilhelm Rontgen, radioactivity by Henri Becquerel and radium by the remarkable Marie Curie, who won Nobel prizes both in Physics and, eight years later, in Chemistry. Curie and her husband worked together on radioactivity and discovered two new elements: polonium and radium. Polonium was named after Marie Curie's home country, Poland, and radium for its intense radioactivity. Both of the elements are extremely dangerous, and polonium ultimately proved to be fatal to one of the Curies. Marie and Pierre's daughter Irene, whose award of the 1935 Nobel Prize for Chemistry made the Curies the most feted family in the Nobel pantheon, was directly exposed to polonium when a sealed vial of the material exploded on her laboratory bench. Irene Joliot-Curie eventually died of leukemia several years later, almost certainly as a result of that accidental exposure.

While radiation in excess can be lethal, we've created increasingly precise ways to focus its beams and manufacture short-acting isotopes. Radioactive isotopes closely related to polonium are used diagnostically and therapeutically in medicine every day; an industry and entire academic fields, such as nuclear medicine and radiation oncology, have evolved from Marie and Pierre Curie's original discovery. We've gradually learned how to safely employ the double-edged sword of radiation in medicine.

In a sense, the problem confronted by traditional radiation therapists and nuclear medicine specialists has always been similar to the one confronted by the Albigensian Crusaders and many armies since, which is how to distinguish between a population of insurgent bad guys, like cancer cells, and the innocents, the normal cells, with whom they are intermingled. In medical terms, the issue is how to cause maximal damage to the target tissue with minimal collateral damage; and we are now developing approaches in which little radiation-emitting "packages," like polonium, are attached to antibodies that will bind only to cancer cells. The radiation precisely given off by the antibody-radionuclide complex will kill the cancer cells and leave the surrounding tissue untouched. In the future, we'll be so precise in our use of radiation that we will be able to act on a cell-by-cell basis, delivering cell-sized doses of radiation to cancer cells one at a time. This highly patient-specific approach to medical treatment is an example of the kind we can expect to see in the dawning era of personalized medicine.

Historically, radiation oncologists have approached tumor-targeting in a way that is very familiar to anyone who has cut a pizza, a pie or a birthday cake. Confronted with a circular pie or cake, most of us will make a series of cuts from the edge through the center of the circle; more experienced cutters will be aware that a typical cake or pie gets pretty chewed up in the middle after a few cuts. Still, this is exactly what the radiation oncologist is looking for in tumor radiation: Beams of radiation are sent from several angles in a circle around a central cancer to cause maximal damage to the tumor in the middle.

The radiation beams are further focused to conform to the silhouette of the tumor as seen from each angle, using metal shields designed to minimize unnecessary damage to the surrounding normal tissue. The shields act like cookie cutters to shape the beam precisely to the tumor. In the 1970s, specific and individually shaped lead cookie-cutting forms were custom-made by skilled technicians for each patient. Today, three dimensional scans like CTs and MRIs and the development of quickly configurable, multipart metal shutters called collimators, which act like metal blinds, make it possible to quickly and automatically shape a beam of radiation to any desired shape. Proton beam irradiation (described in the first chapter) is the latest generation of precision targeting with radiation beams.

The innocent-bystander effect in which normal cells are killed in the process of tumor radiation was brought home to me most clearly before I ever started medical school. My father was a general internist who shared an office for most of his career with his partner, Dr. Sylvan Eisman, who was a cancer specialist. Eisman was fairly short and somewhat stout and one of the early pioneers in the treatment of cancer, a field that evolved during his lifetime to become the specialty of medical oncology. Eisman was an incisive and perceptive clinician and judge of his patients as well as a consummate teacher whose greatest and most charming personal characteristic was his sense of humor. He was just the sort of jolly old soul that Santa Claus is said to be; indeed, Eisman, who was Jewish, dressed up as Old Saint Nick every Christmas Day for over a decade and visited hospital-bound patients with a group of carolers.

My father was tall, thin and reserved, just the opposite of the garrulous Eisman, but however different in appearance, the two of them were role models for generations of physicians who trained with them, myself included. Before I even started medical school, for example, Eisman introduced me to a young woman named Genevieve, one of his patients, about my age, who had been treated with radiation for Hodgkin's lymphoma when she was a teenager. She was very attractive, very engaging and had dealt with this life-threatening illness for more than half of her life. The radiation treatment she had undergone was administered according to the protocols of the era, the 1960s, which called for much higher doses than we use today. She had little tattooed dots on her skin to mark the radiation windows, so that the therapists could align the beams consistently from treatment to treatment.

Genevieve was considered to be a radiation success story, because she was cured of the Hodgkin's. But about a year before I met her, some of the innocent bystander consequences of her earlier radiation began to appear. She had finished college a couple of years before and was working in an office on the second floor of a building. She lived in the city and walked several blocks to work every day, bought a newspaper at the corner and then skipped up the flight of steps to the office. At some point she couldn't precisely pin down, she had begun to develop a sense of heaviness in her chest during the walk every morning, and eventually she began to notice that she had a pain in her left shoulder every morning by the time she got to her desk.

She and Eisman had become close because he had treated her more or less since she was a child. She would make an office appointment and they'd catch up even though the Hodgkin's seemed to be staying cured. At some point during one of these visits, she mentioned this weird shoulder pain, and the chemistry in the little exam room changed subtly but dramatically. Although it had been the usual, uncomplicated banter right up to the point when she mentioned her sore shoulder, the two of them were suddenly engaged in a kaleidoscopic, multilayered fencing match from that point forward. Eisman was a very smart, very experienced, grizzled veteran of diagnostic medicine; his ears perked up like an old dog that caught the scent of something amiss when he heard the words *shoulder pain*. But he loved this girl who had been his patient for a long time and whom he thought of as a daughter, and didn't want to alarm her. Genevieve, on the other hand, while not old, was also an old hand. She had been in a lot of little rooms where bad news was broken, she knew Eisman really well and she had the preternatural, hyperacute sense for trouble common to many long-time patients. He knew something was wrong but wanted to hide it. She knew he had sensed something and wanted to fly. And so they fenced for a bit.

Eisman won this round and eventually ferreted out the fact that she was having pretty regular bouts of chest heaviness and shoulder pain whenever she exerted herself physically. He decided that, while it was extremely unusual in a young woman, the story sounded a lot like angina—the pain from inadequate blood flow to the heart. Genevieve reluctantly agreed to go through the tests, and the studies were unequivocal: She had narrowings in all three of the major vessels to the heart.

The only treatment for coronary artery disease in those days, long before the development of balloons and stents, was surgery—coronary artery bypass grafting—and Genevieve underwent this procedure in the fall of her twenty-fifth year. She was the youngest person the heart surgeon had ever done this operation on. By the time Eisman introduced me to Genevieve, this operation was old news. She had regained her spirits and she now had a pale, asthenic, slightly haunted beauty. Her skin was translucent with subtle dark undertones, and there was a sort of breathless quality to the way she spoke. What I didn't recognize then, but came to, is that the breathlessness wasn't an affectation; she actually had reason to gasp at the end of long sentences. Her lungs were stiff, and like patients with bronchitis or cystic fibrosis, her lips were blue tinged because there wasn't enough oxygen in her bloodstream.

Genevieve had a heart that was fifty years older than it should have been and lungs that were so stiff, she struggled to inhale, because both organs had been injured by the radiation designed to kill the Hodgkin's lymphoma in the lymph nodes in the center of the chest, between the heart and lungs. Lung transplants were not an option in that era, as they are today. Within a year or two of the time I met her, Genevieve died from what was effectively old age. As she neared death, she was still quite beautiful on the outside, but aged and scarred on the inside, a sort of medically created Dorian Gray. As she neared death, her breathing became more and more difficult, and she slowly wasted away as more and more of her energy was diverted to the usually unconscious act of breathing. At the last, she was put on a morphine drip to alleviate some of the agonizing shortness of breath common to people with terminal lung disease. To paraphrase Nietzsche, sometimes "that which kills the thing that does not kill you, kills you." The radiation Genevieve had gotten a decade earlier had cured the Hodgkin's but eventually killed the patient.

When Marie and Pierre Curie discovered radiation, they recognized almost immediately that it had great potential as a medical treatment. Many people still use the terms radiation therapy and Curie therapy interchangeably. Curie actually carried vials of uranium- and radium-containing ores around in her pockets, finding the attractive greenish light given off by the vials appealing. As we know now, the truth about radium and many other treatments is that while they may confer benefits, they almost certainly come with risks. Because we rarely understand all of the risks early on, we seldom know the risk-benefit ratio of a given treatment. In fact, without good outcome data, medical therapies can be very misguided when viewed with hindsight. The application of leeches and other forms of blood-letting for various illnesses, for example, was based on the humoral theory of medicine that prevailed into the late nineteenth century. Humoral medicine has left us with terms originally intended to characterize excesses of one humor or another, like melancholic, bilious, sanguine and choleric. We still use medicinal leeches today to take blood in some unusual circumstances, and we continue to believe that the elimination of what might still be called humors from the blood may improve the outcome of patients with sepsis.

Odd medical practices *still* crop up in strange places. For example, despite the fact that I grew up in the household of a physician, in my youth we put butter on burns as a salve, a practice that probably worsens the outcome. Like the use of magnets, copper bracelets and rhinoceros horns, butter is based on sympathetic theory medicine, another theory that didn't pan out. Both humoral and sympathetic medicines rely to a large extent on the *appearance* of a disease, or what is technically called the *phenotype*. If you are ruddy, perhaps from a fever, humoral theory would have it that you're up a pint or two of blood and should be leeched, or bled. If you have arthritis and your joints are gnarled, sympathetic medicine would say that ground-up shells of a bivalve mollusk known as the Devils toe-nail, and resembling an arthritic joint, are just the thing for you. Feeling flaccid? The Chinese peasant would suggest that a good stiff dose of ground-up rhino's horn will act as an aphrodisiac and stiffen you right up.

Humoral medicine, sympathetic medicine and even today's modern medicine are all based to a large degree on the way a disease *looks* to the naked eye, or even to highly precise, radiation-based studies like CT scans or MRIs. This is to say that much of what we currently know about disease is a function of what's called its *phenotypic* expression or its observable characteristics, which can be distinguished from its *genotypic* underpinnings.

The phenotype is analogous to the skin of an animatronic creature, while the genotype is like the machinery that makes it move. So, for example, while two animatronic King Kongs might make all of the same motions and look identical, phenotypically, one might be powered by hydraulics and the other by gears—they are fundamentally different machines that look the same. Similarly, the fundamental genetic machinery of two prostate cancers that look *exactly* the same but are in two different patients might be very different. As we'll see, a treatment that appears to be eminently sensible based on phenotypic appearance can look very wrong-headed when genotypic information becomes available. It has become increasingly clear that, as we gain more information about the genetics of illness from human genome projects, we'll find that we're going to need to relabel many diseases and rewrite a lot of textbooks.

The humoral, sympathetic and phenotypic eras of medicine are all related in the sense that diseases in each were categorized by very smart, very observant physicians using the tools they had available in their era. Our most modern tools are formidable. CTs and MRIs for example, show us the shape and location of tumors with almost cellular precision. But as with our predecessors, we still don't know a lot about how certain diseases—including many cancers with very precise descriptors—develop and how they eventually acquire the appearance we use to diagnose them with CTs, MRIs or pathology slides. As patients come through our collective medical front doors, we're pretty good at labeling them: "cancer," "trauma," "infection," "diabetes" and so on. But we often don't know exactly why Mr. Jones got his lung cancer other than to observe that he smoked and now has cancer, so we lump him into the smoking-related lung cancer bucket and use the treatments designed for that bucket. We don't really know what exactly broke in Mr. Jones's lungs that started that first cancer cell off in a bad, tumerous direction.

Our confusion about mechanisms is nowhere more apparent than when we call some disease process a *syndrome*. Invariably when this happens, we're pasting a phenotypic label on something that almost certainly is much more complicated genotypically. In the intensive care unit, I regularly deal with two deadly diseases we doctors call adult respiratory distress *syndrome* and sepsis *syndrome*. While the word seems innocuous, alarms should go off in your head when you hear a physician use the word "syndrome."

Syndrome is medical code, and when the doctor uses the term, he is saying, often knowingly: "We've seen enough patients that look like this that we decided to call it a syndrome." But what he means is "We have no idea what causes it; we're not sure how to treat it. And, while we have a theory as to what might be going on medically, to be honest it's not even a little bit complete." He may whisper as an aside, "Between you and me, the drug companies and the government have spent bazillions of dollars, in fact the GDP of small countries, on various studies of this syndrome and search for cures, but nothing really seems to work that well. And I'm very sorry, but the odds are not good."

Syndrome is typically a more fuzzy way of saying phenotype, and we use it as a way of satisfying the desire to cluster things in buckets, to organize our uncertainties in order to deal with them.

If you are of an inquiring bent, you might wonder how we go about identifying a syndrome in the first place. It turns out there is a process in medicine for doing this. Usually a medical professional society or a research organization convenes a committee of Learned Ones to ponder on the definition and treatment of a new syndrome. The Learned Ones congregate in the conference room of some hotel, like a jury, and strive to agree on some definitions and treatments. Their conferences are typically led by a Consensus Builder, who is selected because he or she is slightly less dogmatic than the other Learned Ones, and has a track record of leading Consensus Conferences to a Consensus Decision.

Over the course of two or three days, the Consensus Builder steers all of the other Learned Ones at the Consensus Conference towards Consensus Definitions and Treatments. The first half-day of the meeting is typically dedicated to acronym development. The goal is to discover acronyms that will slip trippingly off the tongue when the *New England Journal of Medicine* or *Lancet* reports study results to the world. For example, the hypothetical VIMCAP (Visually Impaired Males Characterizing A Pachyderm) Conference might spend a day or more dedicated to discussions of and disagreements about trunks, tusks, legs, ears, skin and tails. By the end of the conference, a set of what might be called least-common-denominator observations are codified into a Consensus Definition of the subject. For example, the experts might decide that anything meeting all three of the following criteria fit the consensus definition of a pachyderm: (1) is big; (2) smells funny; (3) moves.

In a process very similar to the one I've just described, experts arrived at the consensus criteria for Adult Respiratory Distress Syndrome, an often fatal lung problem associated with a whole slew of causes, including injury, transfusion and infection. The criteria are: (1) having an abnormal chest x-ray that doesn't look like pneumonia; (2) having problems getting oxygen into the blood; (3) having a heart that seems to be working. While we use the term ARDS to guide treatments, for studies and for insurance payments, we don't really know whether we're talking about one disease or five and how much a person's genetics play into their susceptibility to getting ARDS. That's all about to change as a result of our rapidly expanding ability to match genetics with diseases. And that means that we'll soon be able to precisely tailor treatments to the genetics of individual patients rather than relying on averaged results from large treatment trials of genetically diverse patients.

Like many other industries, the medical profession approaches differences in gender or race-specific treatments or responses to treatment gingerly. Some race-specific diseases are unequivocally genetically based, such as sickle cell disease, while some outcome discrepancies between races or genders may be related only to treatment biases. A recent finding about racially based differences in the way heart failure patients respond to treatment with a beta blocker, a common blood pressure medication, gives us a clue about what the future holds. Beta-blocker treatment has been shown to prolong life in patients with heart failure who are not black, but blacks appeared to do just as well whether or *not* they received a beta blocker. The difference appears to be due to a subtle genetic variant. As many as 40 percent of blacks appear to carry a gene enabling them to make their *own* beta blocker, so they get no additional benefit from a pharmacological drug, while only two percent of whites carry this gene. The same kind of genetic variation probably plays a significant role in the incidence and severity of high blood pressure in different populations. As we learn more about the genetic foundation of disease, the paradigm will change from one in which we treat all heart failure patients with beta blockers to one in which we'll know exactly whether or not an individual patient will benefit. ARDS, heart failure and hypertension are all phenotypes, and phenotypes can be deceiving, as modern biologists have discovered using the ever-increasing amount of information we have from animal and plant species.

The ancient Greeks thought the hippo was most closely related to the horse; more modern scientists concluded that the pig or the peccary were the hippo's more likely cousin candidates based on common, phenotypic characteristics of the teeth. Ridges on the molars of hippos are similar to those of pigs and peccaries. In fact, recent DNA analyses indicate that the hippo's closest living relative is indisputably the whale, and that they share a common, water-loving ancestor that lived about 50 million years ago. The Greeks thought hippos and horses were related because both met the criteria of (1) has legs and (2) eats vegetables. The more recent conclusion clustering hippos and pigs derived from a similar phenotypic error—both species (1) have tusks and (2) have ridges on molars. It would have been a pretty big leap of faith back then to look at a whale, with no legs, and a hippo, with a big distinct head, and conclude that they were cousins, but with genetic information, we now know that the hippo and the whale are very close genetically. Similarly, while two people may have heart failure that looks very comparable phenotypically, the genetic mechanisms and potential treatments appropriate to each might be very different, on a scale like that of the hippo and the pig. Conversely, two individuals may have phenotypically different expressions of a disease that results from a related genotype, and both might then need the *same* treatment—analogous to the hippo and the whale.

With the explosion of genomic information, we are beginning to be able to identify various genes that increase an individual's susceptibility to common diseases like diabetes, heart disease, osteoporosis and back pain. New discoveries are reported weekly, in many cases based on analyses of genetic information from large populations of patients. It is relatively simple to analyze the DNA from a group of prostate cancer patients to determine whether certain gene types cluster within the group. By analyzing these data, we might discover, for example, that prostate cancer patients are much more likely to have one or more specific genetic variants. Alternatively, we might find that prostate cancer patients with a more benign course are genetically different from those with a very aggressive disease. We might even be able to use that information to develop very precise, individually tailored, personalized treatments that act at the molecular or genetic level.

Genetic variants that are more prevalent in prostate cancer patients than normal people have already been identified. It's very likely that a blood-based genetic screening test to identify men at higher risk for the disease will soon be available. Similarly, a genetic test that can be applied to prostate biopsy samples seems to identify more aggressive cancers. And, much as the difference between some blacks and whites in their response to beta blockers during heart failure treatments appears to be genetic, new genetic data seems to explain a similar divide between blacks and whites in prostate cancer. Black men are much more likely to develop prostate cancer and to die from it than white men. For a long time this was thought to be due, in large part, to differences in treatment. More recent evidence suggests that genetic differences actually explain some or all of the difference. The genetic activity in prostate tumor biopsy samples from black men seems to be very different from that of white men. As more and more genetic information becomes available, we'll abandon labels like prostate cancer, ARDS and hypertension and move toward more accurate and informative terms like BRCA-positive prostate cancer that identify the genetic make and model of the disease.

We are entering the era of personalized medicine that has enormous promise but significant peril. On the one hand, we'll all eventually have a copy of our very own genome, and the resulting ability to tailor treatments for a wide variety of potential diseases to our personal genetic makeup—which will be good. On the other hand, there will be tremendous incentives for insurers, employers and perhaps even prospective spouses to have access to that same information so they can better judge how much of a risk we each represent. An insurance company will obviously want to do everything it can to find out which of their clients are at increased risk for the development of expensive, life-shortening diseases. Moreover, there are other, equally predictable, non-health-care-related implications.

Animal species, human and otherwise, instinctively judge the fitness of a mate based to a large extent on phenotypic proxies for good health and good genes, such as the size of a rooster's coxcomb, the brightness of a peacock's feathers and the spread of a moose's antlers, or the clarity of a human's skin, the sheen of the hair and the brightness of the teeth. Of course, we humans game that phenotypically based selection system with tooth brighteners, hair products, high heels and plastic surgery. It's a pretty safe bet that gene screens will find their way into marriage contracts in the very near future, because some people are going to want to know in advance if their intended has a bum set of DNA, however white their teeth or big their pectorals.

Personalized medicine based on genetic transparency has the potential both to inform each of us about our own susceptibilities and to prompt us to do something in response—such as change our lifestyle or environment in an attempt to prevent a disease to which we are predisposed. In pharmacogenomics, an offspring field of personalized medicine, drug regimens and doses can be individually designed for individual patients. Pharmacogenomics is already being used to identify which patients will most *benefit* from certain drugs as well as those patients who may be at special, increased *risk* from certain drugs or drug combinations. For example, several types of chemotherapy are particularly effective in a subset of patients with specific genetic profiles, but more or less in-effective in other patients who have a phenotypically similar disease.

Women who inherit a defective variant of the BRCA gene are at signifi-cantly increased risk for the development of breast cancer, much as men with the same defect appear to be at risk for prostate cancer. BRCA is involved in the self-repair mechanisms for damaged DNA. So BRCA screening can be used for relatives of a breast cancer patient, for example, to determine if they are at increased risk for the development of the disease. Cells have redundant DNA repair systems, and there is a second system that repairs DNA in people with broken BRCA, called PARP (the same poly ADP ribose polymerase I mentioned earlier). While it is clearly bad to have a defective BRCA gene, pharmacogenomics can be used to turn that disadvantage into a treatment for BRCA-positive patients who develop breast cancer. By intentionally knocking out the secondary PARP system, oncologists can render breast cancer cells completely unable to repair themselves.

Both Marie Curie and her daughter, Irene, died of diseases of the bone marrow, both almost certainly related to chronic exposure to radiation. Their bone marrow eventually stopped producing the blood cells necessary for life. Irene died of some form of radiation-related leukemia. Although diagnostic methods were crude when she died, she may have had a chronic rather than an acute leukemia. Chronic myelogenous leukemia, or CML, is one form of the disease that infiltrates the bone marrow and is known to be caused by radiation exposure. Increased rates of CML were seen in Japanese people who survived the atomic bombings at Hiroshima and Nagasaki.

CML is directly linked to a chromosomal, DNA-specific abnormality dis-covered by Dr. Peter Nowell, a faculty member at the University of Pennsylva-nia. Most or all patients with CML have an extra, mutant piece of chromosome, called the Philadelphia chromosome, in the abnormal white cells that proliferate in their bone marrow. These abnormal CML cells produce a special protein that signals the bone marrow to churn out more similarly ab-normal white blood cells. As with many cancers, the abnormal CML stem cells have evolved a mechanism to self-propagate. Eventually, the abnormal cells crowd out normal red, white and clotting stem cells, and CML goes into what's called a blast crisis phase: what was previously a chronic, indolent cancer begins to become very aggressive and deadly.

The very specific, targeted treatment for CML, based on a new, genetically based understanding of the molecular signaling used by the disease, represents a perfect example of personalized medicine. The special, abnormal protein messenger produced by the Philadelphia chromosome can be turned off pretty effectively while CML is still in its chronic phase, thereby blocking progression to the explosive later phases of the disease. A similar, self-perpetuating signaling mechanism is used by other forms of cancer as well, including a different but acute form of childhood leukemia. The same signal-blocking treatment has been effective in these diseases as well. We can expect to see the development of many more very clever genetically based approaches, with treatment regimens that are much more specific and personalized than the traditional, standardized, mass-produced ones used for breast cancer and leukemia.

The kind of genetic profiling that would permit personal prescriptions for lifestyle and drugs probably sounds like something years down the road, but several companies are already offering personal DNA analysis services. For about a thousand dollars and a spit sample, they provide a full analysis of your genome, information about some diseases to which you might be predisposed and even analyze your ancestry. Our ability to decode the genome is only a couple of years old at this point, and it is only the early adopters among us who will get an expensive genetic profile—kind of like buying one of the first personal computers as I did back in the early 1980s. Once you've got it, there isn't really a whole lot you can do with it . . . at first. As with many new technologies, there are a lot of players sitting poised on the sidelines. Once a critical mass of information becomes available about which genes predispose one to what diseases and when, and what to do about it, personal medicine will change health care dramatically and make the current era look like the medicine of fifty years ago looks to us today.

Today's treatments will look like blunt instruments compared to the scalpel-like precision provided by genetic information. Diagnostic DNA-based microchips have already been developed that can analyze a sample of DNA-containing material, such as blood or tumor tissue, and determine the genetic activity in that sample. The large-scale expensive drug trials of the past, that led to the manufacture and marketing of drugs such as Vioxx, with dangerous or fatal side-effects for certain people, will give way to much more specific treatments with drugs that act on the proteins, enzymes and molecules of target cells. The right dose of the right drug will be prescribed based on genetic information about how an individual patient metabolizes that drug, rather than on crude proxies like height or weight. It is even possible that one's own stem cells could be grown in a bottle and have various treatments mixed in to see what works best. Many believe that these targeted therapies have the potential to decrease the overall costs of health care by preventing costly and debilitating diseases like diabetes and heart failure.

One way to think about the current status of genetic information as it relates to the development and treatment of diseases is to imagine that our ultimate goal is to make a full-length, color, three-dimensional animated movie, describing the life and adventures of genes starting at their very first moments in the zygotic mother of all stem cells. Today, we only have a few partial sketches of the storyboard that will eventually outline the movie. We need much, much more information about lots of people, with lots of diseases to fill out all of the frames in the storyboard. To color and animate the whole movie we need a lot more information about the behavior of lots of diseases as they evolve over time.

Dramatic advances in our understanding of the molecular mechanisms of many common diseases are occurring in laboratories all over the world every day—filling in those sketches, as it were. And it won't be long before the development of very specific, genetically based, targeted treatments for diabetes, Alzheimer's disease, heart disease, cancers and many other illnesses. The era of blunt-force treatments, like the radiation Genevieve received for her lymphoma, is drawing to a close, and personalized, genetically based medicine is in its infancy. Perhaps, if some optimists are correct, someday we'll understand the genome so well that we can even design interventions to turn off the genetic aging process, and our film will be titled *The Never-Ending Story*.

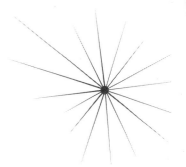

CHAPTER 13

MY FATHER'S FEET

In the class on computers in medicine that I taught for several years at Princeton University, one of the assignments I gave the students was to imagine ways of embedding health care into their home. The intent of the assignment was to force them to think about ways to extend the perimeter of health care beyond the doctor's office and the hospital, right into the home. The range of responses I got over the years was what you might expect from a smart, creative, technologically sophisticated group of college students. Some described calorie-savvy refrigerators, others novel health sensors. A surprising number came up with toilets that could monitor body fat, weight or excretions. A majority of the students envisaged some kind of on-demand Internet link with a health care provider.

Princeton is close to Philadelphia, and most days I would drive up to the university to teach and return to the hospital later in the day. The transition from the bucolic, spacious Princeton campus to the busy, urban hospital was invariably jarring. The drive itself was stressful, because I had to travel along one of the busiest highways in the United States—Interstate 95—which goes from Maine to Florida. The faculty and students at Princeton were relaxed and quiet. The faculty, students, employees and patients at the Hospital of the University of Pennsylvania were hurried and stressed. Princeton was cerebral, Penn utilitarian. On a bad day at Princeton, the slide projector malfunctioned. On a bad day at Penn, someone died.

Curiously, when I would ask the anesthesia and surgical trainees I was teaching at Penn to answer the same questions I had just left the Princeton students with, they came up with different answers. To them, the act of touching their patients, sometimes intimately, was the way one practiced medicine, while the Princeton students had no problem with the idea that they might one day communicate with their doctor over a Webcam. Of course, back in the old days, when my father practiced, physicians *were* available, on demand, and in one's home just as the Princeton students imagined. And when the doctor came to your home, he brought a little black bag and the most sophisticated medical tool available at the time, his hands.

My father's hands were large, warm, blunt-tipped and always well manicured. He was an internist and used his hands as diagnostic tools. His fingers probed his patient's neck, abdomen, armpits and groin—the soft underbelly as it were—searching for enlarged nodes or organs. These are private parts and a patient is most comfortable when the examiner's hands are warm, clean, well-groomed and competent. My father knew this. He was reserved, and many people described him after his death as "a gentleman." He shook my hand, in what must have been a very proud moment, as I was about to be inducted into the Philadelphia College of Physicians, an old honorary society dating back to the 1700s of which he was a member. He said to me quietly, "Your hands are warm, you're not nervous. That's good."

As a child, with fascination I used to watch him trim, file and clean his nails. He even applied something called cuticle wax during one period, which was not, I think, an expression of personal vanity, because he wasn't that way. It was done more in the line of a master craftsman's maintenance of his tools, much as the carpenter would sharpen his planes, or a butcher hone his knives to a razor's edge, with a drop of oil on a whetstone.

Physicians are taught to inspect, palpate, percuss and auscultate, or look, touch, tap and listen, and as a routine stage in the physical examination of a patient, my father would pass his hand under a patient's garment to feel the character and location of the beat of the heart within the rib-cage. In order to feel the apex of a woman's heart, he would have slid his right hand under her left breast, exactly as I later learned to do.

He touched with the tips and sides of his right index and third fingers—even a left-handed physician, as he was, examines the patient from the right side because that's the most natural way to feel the heart. When percussing, he used the middle finger of his dominant left hand to strike the last knuckle of the middle finger of his right over the patient's chest and stomach, acquiring information from the resulting sounds. I use my right hand to strike my left because I'm right-handed. The striking finger is called the *plexor* and the struck finger is the *pleximeter*. Done skillfully, percussive examination can tell the practitioner a lot about the character of the tissues and organs that lie beneath. A stomach full of gas sounds like a kettle drum. A healthy lung has a distant but still-hollow sound. The lung of a patient with emphysema sounds more resonant than normal, but a collapsed lung sounds dull. The liver sounds flat, more like muscle. An experienced examiner can map out the location and determine the character of major organs without need of an x-ray. It took me some time to become competent at percussion—the plexor finger must be kept rigid as the hand is snapped, rotating quickly through one of the wrist's degrees of freedom toward the pleximeter finger of the opposite hand. The wrist stays loose and the finger firm. Unfortunately, physical examination is an increasingly lost art in the medical schools of today.

My father often used those same big blunt, skilled hands to examine painful areas, such as the belly of a patient with appendicitis or an inflamed gall bladder. Here, a doctor must strike a balance between touching firmly enough to gather necessary information and gently enough to minimize discomfort. A good physician improves the calibration of his hands as he goes through his career, touching, assessing, integrating and touching again.

All animals use touch to explore their world, and touch may be the most primitive sense. Parents feed, carry, teach and discipline their offspring using various modes of touch. Touch is important during courtship and coupling. Primates groom one another with their hands to reduce stress, win favors, avoid confrontation and establish or reinforce social hierarchy. We humans run to be touched by our parents, we hold hands, we go to barbers, manicurists and masseuses. As far as I know, however, humans are the only species that touch one another to diagnose and treat disease.

I have a medical tool on my desk today that I don't use very much. It is the sphygmomanometer that my father carried, and its name derives from the Greek roots for the words for pulse *(sphygmo)* and sparse *(manos)*. The sphygmomanometer was designed to measure something sparse in the pulse—life, perhaps. The beautiful, wood-cased, hinged instrument on my desk consists of a glass column with a bubble reservoir of mercury at the bottom, very much like a thermometer. The column is attached by rubber tubing to a cuff that is designed to be wrapped around the arm, and there is a compressible pump-bulb attached to the cuff. The technique for measuring blood pressure is simple. The doctor squeezes the bulb, which pumps air into the cuff, gradually tightening it around the patient's arm. As the pressure increases in the cuff, the mercury climbs in the tube, measuring the pressure in millimeters of mercury. When the cuff pressure is high enough, blood stops flowing in the limb around which the cuff is wrapped, and the sparse pulse of blood flow below the cuff disappears.

Once inflated, the cuff is gradually deflated by releasing air through a screw valve, and the mercury slowly falls in the glass while the physician listens with his stethoscope to what are called the Korotkoff sounds over an artery below the cuff. Nikolai Korotkoff was a Russian surgeon who identified the audible pulse sounds as those of a "passage of part of the pulse wave under the cuff." The wooden case of my heirloom sphygmomanometer is mahogany, handmade and well worn, and the fittings are brass. It is still a functional device and fine to look at, but today it is a beautiful, if somewhat archaic, piece of medical equipment. Today, you can walk into a drugstore and stick your arm into an automated blood pressure measurement device that uses this same technology you'll soon be able to buy as one of your home's network-enabled health-monitoring devices.

Sphygmomanometers are like many signature pieces of medical equipment, such as the otorhinolaryngologist's head-mounted mirror or the anesthesiologist's laryngoscope. The function and working of these tools is not intuitive, and one can instantaneously distinguish between the skilled practitioner and the novice by the way they handle the instrument. I watched my father use his wood-cased sphygmomanometer, which was labeled "Lifetime Baumanometer" and patented in 1924, when I visited his office as a child. He would deftly pull up a patient's sleeve and wrap the cuff around an arm. His free hand reached back automatically for the stethoscope that was always just so in the left-hand pocket of his starched white jacket. With the skilled physician's version of a quick-draw, he donned the instrument with one hand, his fingers spreading its earpieces just wide enough for his head, and inserted it into his ears. He used his right hand to hold the bell of the scope over the artery in the crook of the patient's elbow just below the cuff, while with the left he pumped up and deflated the cuff, watching the mercury fall in the glass tube and noting the Korotkoff sounds as they came and went.

His head was bent down toward the patient, partly because the tubing of the scope was short enough that he had to, and partly because that was the way skilled doctors examined—in close. He kept one hand in contact with the patient at almost all times throughout the physical exam. Many good doctors do this, both as an act of reassurance and a subtle indication of their authority to touch and examine.

Medicine has changed dramatically since the days I watched my father practice. House calls are a thing of a vanished era, except perhaps in boutique practices where an internist contracts to be available in that old-fashioned way to persons of means who are willing to pay a premium for extra service. There are a host of explanations for the obsolescence of house calls: they are inherently inefficient, the geographic spread of practices has increased dramatically, there is no reimbursement commensurate with the time involved, and homes lack the diagnostic equipment required today—hands are no longer sufficient. The pendulum is about to swing, however, because of technological advances and changing health care economics.

When asked to imagine home-based health care in the future, most of my computer-literate students at Princeton imagined some kind of intelligent system in which multiple smart, medically aware appliances—perhaps the refrigerator, perhaps the toilet—monitor the home's occupants. They envisioned software that would gather information from these devices and package it into a coherent format. Some even imagined a smart house that monitored the movements, caloric expenditures and vital signs of the occupants. They all assumed that there would be digital medical diagnostic tools of one sort or another in the home.

It turns out that the imaginations of these relatively health-care-naïve students come very close to what we are about to see with the blossoming of "home telehealth." There are already digital sphygmomanometers and blood glucose analyzers, digital motion sensors and scales, digital electrocardiographic equipment and digital pillboxes. There are also a number of useful medical functions that could be integrated into refrigerators and toilets. A refrigerator could very easily keep track of calories and medications; a toilet could be designed to measure weight and body fat and to sample the urine for medically relevant chemicals. Many of the currently available medical monitors come with a plug and a digital display, but only a few are capable of being plugged into a computer. Software giants like Microsoft and Google are already thinking about ways to gather information from the next generation of home medical devices.

As the college students in my class presented their projects, many of them told stories about medical illnesses they'd experienced personally or witnessed in friends and family members. These students seemed particularly attuned to the vulnerability of someone who is chronically ill. Some had sick relatives who required constant support from family or nurses; others had elderly, frail parents who lived alone. A few of the students seemed to understand some of the barriers inherent in a health care system that requires the patient to get to the doctor, rather than the other way around.

The house call of yore went something like this. Someone got sick, so you rang up the doctor, and he packed up his equipment and drove off to your house. He examined, diagnosed and prescribed for your problem using the tools in his bag. The encounter might take an hour or more of his time. You paid him with a pig, or some preserves, which was a pretty reasonable exchange of services at the time. Today, however, the insurance reimbursement to a physician for that same encounter would be a good deal less than what the plumber, appliance repairer or a computer geek who makes house calls would command for the same visit. So instead, patients come, often at some significant inconvenience, to the doctor, wait in the waiting room, sometimes for unconscionable periods of time, and, when their turn comes, are then hustled through their appointment. The house call gave the patient the home field advantage, while today the doctor clearly has the edge. Telehealth will level the playing field.

Justin Starren, a physician friend of mine, is a member of the faculty of the Department of Biomedical Informatics at Columbia University. I invited him to give a guest lecture to my Princeton class a couple of years ago. Justin began the class by describing a problem he and some colleagues were trying to help solve in New York state. Not surprisingly, there are a huge number of elderly ethnically diverse patients with diabetes all over the state, many of whom get marginal medical care for a variety of reasons. Some are so disabled that they have trouble getting to and from the doctor's office; others can't afford the cab fare. Some have trouble understanding how to manage their disease; others have language barriers that interfere with their ability to ask the questions they need to ask in order to participate in their own care. Starren and his colleagues decided to see if telehealth might provide a solution.

They designed a study in which they compared a group of patients treated routinely—that is, with regularly scheduled doctors' visits—to another, randomly selected group of patients managed with telehealth. Both groups of patients had the following characteristics: They were older than 55, on Medicare, had been diagnosed with diabetes by a doctor, were treated for it and lived in a federally designated, medically underserved area. They had to speak English or Spanish.

The telehealth patients were given a home telehealth unit that consisted of a computer connected to a telephone line, a digital glucose and blood pressure monitor, as well as access to an educational Web site with normal- and low-literacy versions, in both Spanish and English, of a diabetes educational tool. About 800 patients were enrolled in each group, and only about 20 percent of the study patients started with any idea of how to use a computer. The telehealth group was taught how to keep track of their blood pressure and glucose levels on the computer, upload the data to the study center and initiate a videoconference with nurse case managers. At the end of a year, the glucose management, blood pressure and cholesterol levels of the telehealth patients was significantly better than that of the patients managed with usual care. In other words, not only was it more *convenient* for the telehealth patients, they had *better* outcomes.

The first study was so successful that a new version of the study has been underwritten by the U.S. government with some minor modifications. New customized telehealth computers were issued to study participants. They have a very simple interface device with three buttons labeled in English and Spanish. One push of one button initiates a video-conference with a remote nurse manager. Another button sends the blood sugar and pressure data to the Columbia web site. A third brings up the Web-based instruction page. The computer mouse turned out to be a problem in the first study because of unexpected things such as the size of the pointer, which was difficult to see for some patients, and the degree of coordination required to point and click, which was beyond some patients with nerve disease from the diabetes. It is not so simple to point and click correctly when you have a tremor, for example, or when your fingertips are numb. The new devices also have a restart button, a simple way to start over, something those of us who've talked an older relative through a computer problem would appreciate. Finally, the new computers have a video camera with which the patient can photograph her hands or feet if requested, so that the doctors can see if there are lesions that might need an in-person evaluation. The blood pressure and glucose data are analyzed by the computer as the data are acquired in the home, and a provider is automatically notified if they fall outside of certain safe bounds. Of course, all of the transactions are conducted over secure interfaces to preserve patient privacy, in the same way that we protect financial transactions over the Internet.

One of the obvious questions about a study like this is whether the expense of the necessary computer infrastructure exceeds that of usual care. The answer is pretty clear. Diabetes is only one of many similar chronic medical conditions. Diabetes, high blood pressure, obesity and smoking are diseases where the downstream costs accrue like compound interest in a bank account. The tighter glucose is managed on a day-to-day, hour-to-hour, even minute-to-minute basis, the healthier the patient in the long term. Direct complications from diabetes cost 45 billion dollars per year in the United States and an additional 45 billion in indirect costs come from diabetes-related illnesses—that's the compound interest. So any immediate savings from telehealth, even with the cost of the equipment factored in, is likely to result in great long-term savings to the health care industry, not to mention the convenience factor for the patients who no longer have to make routine trips to the doctor.

The patients in the Columbia study have their simple, specially designed computer propped up on some multipurpose surface in their home. Maybe it's a kitchen counter, perhaps it's a dresser in the bedroom, or maybe even an heirloom side table originally crafted by an ancestor of the patient in some other country. Digital home health equipment in the coming decade, however, will be highly sophisticated, artificially intelligent and integrated into the furniture; it will also be only a part of the rest of the home network, along with the telephone, television and home computing system. If you have trouble conceiving of what this might look like, consider the smart refrigerator that knows what it contains and can automatically order replacement food when needed. LCD screens, televisions and computer connections are being integrated into newly built bathrooms by a number of manufacturers. You can buy digital thermometers, blood pressure cuffs, and glucose analyzers at your neighborhood pharmacy. The Japanese manufacturer DoCoMo has even developed a cell phone that can track how many steps you've taken each day, your pulse and, if you breathe into it, whether you have bad breath. These are all examples of the kind of intelligent, health-aware infrastructure that will eventually be part of your home.

My father never owned a cell phone, never had a computer in his home, wrote his patient notes in an almost-illegible hand in patient charts he kept in file cabinets in his office, and he made house calls. By way of contrast, the next generation of internists will never be without a cell phone, won't be able to practice without a computer, will dictate their notes to a computer transcriptionist and will make lots of virtual house calls every day. A couple of things will need to change, however, before that happens. The first is that doctors will need to be paid for telemedical encounters. Reimbursement rules for various types of medical encounters are confusing and contradictory. For example, most insurers don't pay for patient-physician e-mail, but online consultation may be paid for. This barrier will go away eventually because the drivers for telemedicine are so great, including improved efficiency and patient convenience. In fact, one of the government's interests in the Columbia University diabetes project is in using it to determine how to pay for teleheatlth.

The Columbia project provides another clue about how health care in your home is likely to work. The second generation diabetes management computer has a way to tell if it needs to notify a provider about problematic data. Just like the telemedical ICU that I described earlier, the Columbia diabetes program has a background intelligent alarm system built into it. Home telehealth care systems of the future will have several types of built-in, health-oriented software. There will be a function that identifies problem trends in your weight, blood pressure, glucose, eating habits, heart rate or even more esoteric indicators. This function may be programmed by you to notify your doctor automatically or perhaps you'll choose to keep the information to yourself.

There will also be something that acts like the software wizards provided with many software packages today. You will use this function to troubleshoot routine medical problems. Imagine the following dialogue between you and the automated health care wizard you've summoned up like a genie after your child cut her finger on a piece of glass:

Wizard: Is the wound dirty?
You: No.
Wizard: Is the wound still bleeding?
You: No.
Wizard: Is the most recent tetanus shot within 5 years?
You: No.
Wizard: Is the wound longer than 2 centimeters?
You: No.
Wizard: What is the location of the wound?
You: Finger.
Wizard: The wound does not require suturing, a Band-aid or tissue adhesive can be applied per manufacturer's directions. A tetanus booster is indicated.

Simple software like this can be used to guide many people through relatively straightforward medical problems and preempt unnecessary medical visits.

A third, critical health care software function will be a personal medical information file. We will all eventually have access to and responsibility for a large portion of our own medical records. While the privacy issues are formidable, personal, on-line medical records are just around the corner. Microsoft and Google are both positioning themselves to provide us with secure on-line medical storage. Google Health offers a medical record that allows you to enter conditions and symptoms, medications, allergies, surgeries and procedures, test results and immunizations. Microsoft's HealthVault is designed to be a free, on-line encrypted data vault for health information. And, unlike the Google offering, HealthVault can be configured to automatically acquire data from a variety of home digital medical equipment, including blood glucose, blood pressure and breathing monitoring equipment.

Whether through Google, Microsoft or some other large vendor, perhaps even the government, you will eventually keep your medical data in some secure on-line location to which you grant access to specific outside providers under specific rules of engagement. You might choose to grant broad access to a relative or health care proxy. You may let a pharmacy company maintain your medication records and link them to your medical record. Your doctor or a hospital may have the rights to update certain parts of your record. The medical equipment in your house may be able to automatically write new information into your record.

Home telehealth is in its infancy, but its evolution is likely to parallel the one that we've seen in banking over the past decade. Twenty years ago, I would be issued an envelope with a paycheck at the end of a week, take it to the bank some days later for deposit, write checks to pay my bills and use traveler's checks any time I went farther than a couple of hundred miles from home. Today, my paycheck is deposited automatically; automatic deductions go from my account to fund college savings accounts and the like. My bills are paid automatically according to specific rules I've set. I can go to Europe with twenty dollars in my pocket and expect to find an automated teller machine at my destination that dispenses euros. And I get secure, automated e-mail keeping me up to date on all of this activity. I can access the bank 24 hours a day to review my finances, move money around and get financial advice. Home telehealth will get to this same level of freedom and flexibility in a lot less time than banking did.

Medicine is moving from the hallowed ground of the hospital and the doctor's office into the sacrosanct rooms of your home. We're transitioning from the medical world of my father, in which medical information was secret and closely held by medical high priests, to one in which there are medically oriented search engines on the Web that can lead you to information even your doctor doesn't know. While your medical record was once kept in crypts in the hospital, and, at least in my father's case, recorded by doctors in what might have been Sanskrit, you'll soon own and maintain your own digitized medical record on the Web. My father's black bag contained medical instruments that, over the course of a lifetime, he learned to use like a master craftsman. The contents of *your* black bag will be purchased at a pharmacy, come with a user's manual and interface wirelessly to your home network.

We're all going to have access to the tools, the information and the responsibility to take better care of ourselves. If I'm your insurer, and I know that an elderly, computer-illiterate, diabetic new immigrant can maintain very tight control over her blood glucose with the help of home telehealth, I'm probably going to penalize you for mismanaging your own diabetes provided you have the same tools. Eventually, we're all going to have to figure out how to operate in this new world and to find more cost-effective ways of providing health care. As doctors, we'll need to learn when and how to integrate virtual visits into your medical care and when we need to see you in the flesh. As patients, we'll have to take on greater ownership for our own health.

In my experience, many patients will embrace advances in technology that help them navigate through our enormously complex health care world. As a friend of mine put it after she and her husband went through several complicated medical encounters, "We need guides!" She said, "We're well educated, we're young and we have means, and yet we couldn't always get the medical help we needed." Their frustrations arose in their attempts to help care for parents and friends who were older and had become ill. They were fielding seemingly straightforward questions to which they couldn't easily find answers. The questions were as simple as, "Is this normal?" "Do I need to see the doctor for this?" "Is there anything I can take for this?" or "How is this likely to play out?" With all of our increasing technological sophistication, we've lost some of the magic that people like my father's partner, Dr. Sylvan Eisman, and my father himself brought to the care of their patients. They knew people. They had and took the time to listen for the little clues and anxieties that many patients brought to their offices.

I distinctly remember my father telling me about a first office visit with a female patient that he felt had gone poorly. Over the years I've heard from many patients about how check-ups with my father would invariably include a period when he sat at his desk across from the patient and had a free-form, non-goal-directed chat. I know he learned a lot about medical issues and a lot about the people he cared for that way. But this particular visit hadn't gone well, and after mulling it over, Dad concluded that it went wrong the moment he propped his feet up on the desk and leaned back in the chair to listen.

She subsequently became a long-term, devoted patient, but Dad felt he'd stepped too far inside her personal space on that first visit. Of course, I'd bet that there are a lot of us today who would love to have unpressured time to talk about important things with our doctor while still fully clothed. It is very unlikely that your doctor will ever throw his feet up on your dining room table during a leisurely interview, but there is every reason to believe that you'll be able to talk comfortably and confidentially with him from your home over a live, two-way network link in the very near future. There's no reason to think that we can't combine the benefits of advancing medical technology with the humanity that's always been a fundamental part of medicine.

My father died one night in his sleep over fifteen years ago. He had worked all night in an emergency room the night before, and gone to a party on the night he died; I got a shattering phone call from my mother in the middle of that night. Recently, I got another phone call like that, from my brother, when I was on a vacation in Costa Rica; this one was to say that my mother, who is a fiercely independent woman in her late seventies, had slipped on a patch of ice while taking out the trash. She'd lived alone since the night my father died and planned to continue doing so as long as her health permitted. My mother, who was alone at the time, had fallen and couldn't get up because it hurt too much—and it was below freezing.

Serendipitously, a friend arrived within minutes and, with some difficulty, helped get my mother into her car and off to a nearby clinic. X-rays showed that she had broken her pelvis, fortunately in such a way that she didn't need surgery. At this point I got the news, thousands of miles and a time zone away. I am the only physician in my family, but my siblings and I were all running through the same series of thoughts: "What was she thinking, walking on

ICE!" "Thank God she didn't hit her head." "Thank God she didn't break a hip." "What do we do now?" She didn't need to be in a hospital, although she certainly could have been admitted somewhere given the circumstances. And she wanted to stay in her home despite the fact that she couldn't walk, couldn't cook and couldn't get up the steps to her bedroom without help.

After a lot of thought and a lot of phone calls, we finally figured out that this would probably be all right. She figured out how to go up the stairs safely, one by one, sitting, pushing up a painful step at a time. We brought in food. We hung a phone around her neck. She took pain drugs and aspirin to prevent clots from forming in her legs. Eventually she was well enough to progress from using a four-legged walker to a cane with the help of a physical therapist who came by regularly. She didn't answer the phone immediately when we called a couple of times, which was alarming, but this was usually because she was talking to one of her many other friends at the time. She made that first clinic visit and just one doctor's visit during the entire two-month saga, and is exercising on a treadmill today, just like she did before the accident—back to normal.

The success with which she managed to navigate this major health problem and still manage to stay in the comfort of her own home was in large part due to her proud spirit and fortitude. My siblings and our spouses were able to help, and I was able to get answers to such questions as: "Is this normal?" "Do I need to see the doctor for this?" "Is there anything I can take for this?" or "How is this likely to play out?" But, were the new systems in place, she could have done this without us. There are compelling reasons to develop and invest in technologies, like the ones used by the diabetic patients in the Columbia study, that will allow people to prevent disease, manage chronic illness and navigate health care crises with as much self-reliance as possible.

I talked to my mother about her experience the other night while I was driving home from work. A routine conversation unexpectedly veered off onto a philosophical tangent, and I know my father sent a lot of conversations off the same way by putting his feet up on the desk. My mother, who doesn't ordinarily like to talk about this kind of thing, said: "I've watched a lot of my friends die recently, and I've come to realize that you're pretty fragile when you're my age. You can be healthy one day, and the next you're really sick . . . or dead." We both knew she was thinking of her fall. Then she said: "But there you are! And I'll see you in a week when I get back from my *fabulous* trip to Florida." And I smiled.

The odd thing about that initially casual phone call that turned into a really important conversation, one that she and I might never have had in person, was that I was talking to her on a visor-mounted, Bluetooth-enabled speakerphone, wirelessly linked to my Blackberry smart phone as I buzzed along at seventy miles per hour. Some fear technology, while others hide behind it, but done right, technical and medical advances will not rob us of our humanity. I *know,* for example, that Stephen Hawking has a biting wit, despite the fact that he communicates through a text parser and a voice synthesizer, just as I *heard* my mother's wistfulness and courage through the complicated hardware mounted in my car. Distilled to its essence, medicine is the instinctive desire of humans to reach out to and care for others, and that instinct must be our pilot as we sail off into the sea of possibilities ahead.

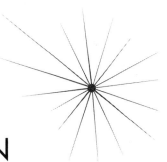

CONCLUSION

A lot of things are going to change very rapidly in the way we practice medicine over the next few years. As with many other industries, the pace of change is already accelerating rapidly in comparison to previous decades and eras, and the evolutionary pressures are irresistible. We spend ever-increasing amounts of resources on health care with apparently diminishing returns. The watering hole is getting drier. At the same time, the promises inherent in many of the technologies I've described in this book will be irresistible—personalized drug treatments, cures for baldness, organ rejuvenation, suspended animation and perhaps even perpetual youth.

Some of the treatments and technologies I've described will fall under the straightforward rubric of medical treatment, and therefore be handled in the same way diseases are in most countries today, with some reasonably democratic distribution of the available treatments. In other words, in most countries you can't currently buy better medical care. There are reasonably crisp edges between treatments for disease that are covered by insurers and discretionary medical procedures for cosmetic purposes for which the buyer pays out of pocket. To be sure, there are hybrid interventions, like in vitro fertilization, that are handled very differently in different places, but most people have a pretty good idea of what insurance should cover and what's discretionary.

That's all going to change in the near future. It's already possible for people of means, the haves, to buy their own genetic profile or that of their entire family. Similarly, parents can purchase cord blood storage for their children if they have the means. Cryonic preservation, whether or not it will ever result in one restored life, is an option available only to the very rich. And in the very near future, we will have developed stem-cell-based regenerative treatments that will permanently blur the boundary between therapeutic and life-extending interventions. For example, the exact same stem cell treatment might be used to treat a pathological kidney disease or solely to regenerate an aging kidney. It will be very difficult to draw a line dividing medically indicated kidney stem cell treatment from elective, life-extending treatment.

The irresistible lure of performance-enhancing interventions for athletes, even high school students, gives us a hint of what may lie ahead. The potential gap between the haves and the have-nots will only widen in the future, as performance-enhancing neural implants become available. The day *will* come when it will be possible for a person with the means to buy a neutrally implanted network interface or prosthetic eyes and ears that confer supernatural senses on the owner.

It's hard to imagine the consequences if we ever identified some genetic, fountain-of-youth switch that turned off the aging process. Would it be covered by health insurance? Would it be something one could buy? How much would it cost? And what would happen to the planet should it ever become available and people, or perhaps only some people, no longer died of old age?

I did not write this book as a bright-eyed endorsement of medical technology. As I walk through the operating rooms and intensive care units of our hospitals, I see many patients whose lives were extended by our technological capabilities, our efforts often resulting only in extended, sometimes painful, disability with no improvement in quality of life. On the other hand, each of the technologies I've described has the potential, if applied thoughtfully, to improve the quality of our lives and reduce the overall costs of health care. Moreover, these benefits could be realized within our own lifetimes.

In my view, we all have an increasing responsibility for our own individual health as it becomes ever clearer how much of a role personal behaviors play in disease and aging. I also believe that as a society, we have a responsibility to create incentives and penalties that promote healthier lifestyles. Those of us who work in health care have our own responsibilities as well. If physicians of previous generations were overly paternalistic, in some cases making decisions for patients without their participation, the pendulum has swung back the other way. It is in many ways easier today for a physician to justify doing everything, and thereby avoid difficult conversations and legal tangles, when doing less might be appropriate. More importantly, the health care industry will need to deploy new technologies with close attention to their cost implications. In other words, we can't just slap another layer of paint on top of the old one; we have to reengineer current processes as new technologies become available and eliminate waste in the process. In some ways, today's health care performs like an old computer getting more and more inefficient with each added program.

More than ever before, medical advances are inextricably linked to advances in other industries. New developments in computing, such as networking, data visualization and artificial intelligence, have immediate and direct applications in medical care and treatment. Robots are increasingly deployed in manufacturing, aviation and the military, and there are a wide variety of potential medical applications for these versatile devices. Genomics will play a role in medicine, agriculture and conservation. Molecular, nanoscale machines will proliferate as the technology evolves and the potential applications for tiny, novel biomechanical devices are limited only by our imagination. And, as with the NASA program, developments primarily designed for medical application will generate spin-offs extending outside the scope of their original intent, perhaps even a direct connection between man and computer.

The other day I was walking down the hallway outside the intensive care unit when I came across a six-year-old boy scampering around energetically. His mother was visiting the boy's father who had just undergone brain surgery. As I walked by, I saw Tug, our little robotic pharmacy courier, turn the corner and head toward the boy. I stopped for a second, curious to see what would happen. Tug pulled up to a stop in front of the boy as he was programmed to do in the face of an obstacle. Tug and the boy were just about the same height, and the little human cocked his head for a second, never having seen anything like the robot. There was a momentary pause as they sized each other up. Tug blinked first and said "Waiting to proceed," in his flat, mechanical voice. The startled boy hopped out of the way and ran to his mother as Tug sailed off down the corridor.

I thought to myself, "We're not waiting to proceed in medicine. Not even for a nanosecond."

INDEX

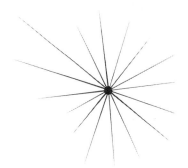